EXPL**O**DAPEDIA

KW-480-388

Welcome to *Explodapedia*, the indispensable guide to
everything you need to know!
This series is packed with in-depth knowledge you can trust;
it gives you the tools you need to understand the science
behind the wonders of our world. Read on to learn all about
our miraculous minds in *The Brain* . . .

'*The Brain* is splendid - fun to read and really exciting.'
Henry Marsh CBE, neurosurgeon

'Elegantly balanced between soaring possibility
and informative realism'
Guardian

'Extraordinary discoveries are explained in this book in
a way everyone can understand.'
Sir Paul Nurse, Nobel Prize winner

'The perfect balance between charm, quirkiness and
wonder . . . for kids and adults alike.'
Siddhartha Mukherjee, Pulitzer Prize winner

'A totally fascinating book, brimming with amazing scientific
knowledge and fab illustrations.'
Greg Jenner

EXPL☆DAPEDIA

THE BRAIN

What Goes On Inside Your Head?

Ben Martynoga

Illustrated by
Moose Allain

FICKLING
db
David Fickling Books

Explodapedia: *The Brain*
is a
DAVID FICKLING BOOK

First published in Great Britain in 2025 by
David Fickling Books,
31 Beaumont Street,
Oxford, OX1 2NP

978-1-78845-274-8

1 3 5 7 9 10 8 6 4 2

Papers used by David Fickling Books are from well-managed forests and other
responsible sources.

DAVID FICKLING BOOKS Reg. No. 8340307

A CIP catalogue record for this book is available from the British Library.

Printed and bound in Great Britain by Clays, Ltd, Elcograf S.p.A.

Italic type is used in Explodapedia to highlight words that are defined
in the glossary when they first appear, to show quoted material and
the names of published works. Bold type is used for emphasis.

Contents

A Whole Universe Inside Your Skull

Picture the scene: you're finally knuckling down to finish the project you've been putting off all week, it's due in tomorrow, and you've just about got time to do a decent piece of work.

You open your laptop and start tapping away when – 'bing' – a message arrives.

It's from your friend Ash: 'Amazing! Got returned tickets for the gig tonight! Come with us!!!'

Oh wow! You tried everything to see this band, but tickets sold out months ago.

. . . But . . . Why tonight?

It's decision time. Do you play it safe, stay home and study, or seize the moment and head to the concert?

Chances are you've now got a heated debate starting up inside your head, with different 'voices' yelling things like:

It's noisy in there, right?! But where the heck do those vivid sensations and yabbering voices actually come from?

Your brain, of course. We'll return to this conversation later, to find out exactly how – and why – your brain can end up arguing with **itself**. But right now, the key thing you need to know is that your skull contains one of the most complicated, creative, awe-inspiring and bamboozlingly powerful objects in the universe!

That said, to be honest, the human brain is kind of unimpressive to look at. Roughly the size and shape of a large cauliflower, it's a dull greyish pink in colour. And if you jabbed it with your finger, it'd be like prodding a massive soft-boiled egg.

But, somehow, that fleshy-pink, bulging, slightly wibbly cauliflower is the most essential, irreplaceable part of your entire being! As we'll see, it's in control of almost every part of your body and practically everything you **think**, **feel** and **do**. It can make your mouth water at the thought of a biscuit. And, at the same time, thanks to its amazing powers of memory and imagination, your brain can turn itself into the best time-travel machine ever invented.

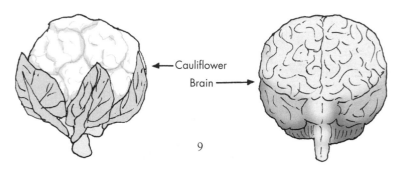

Cauliflower ← → Brain

So if, for any reason, you're feeling a bit stupid or down, or you're just having a rubbish day, try to hang on to this fact: your brain is one of the brightest stars in the universe. In a way, it actually **contains** the entire the universe.

In fact, it was a poet called Emily Dickinson, not a brain scientist, who first pointed this out, way back in 1862:

Eh?

The Brain—is wider than the Sky—
For—put them side by side—
The one the other will contain
With ease—and You—beside

What she meant was that absolutely **anything** and **everything** that you can sense, feel or even just **imagine** must, somehow, exist **inside** your brain. And that includes the universe itself! Since you can **think** of it as a 'thing' – an almost unbelievably massive one, sure – it must be able to fit inside your skull.

Wow!

Not bad for a wibbly, greyish-pink cauliflower, huh?

Yeah, yeah, yeah. Blah, blah, blah. Yadda, yadda, yadda.

Um . . . hi, Octopus. Didn't realize you could, er, talk . . .

You lot aren't the only intelligent beings on this planet, you know.

We-ell, we do have the most complicated brains in the animal kingdom.

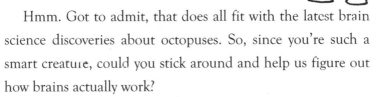

Whatevs. You've only got one brain, though. We've got nine.

Really?!

We use our brains for learning, communicating, using tools, cracking puzzles and generally having a blast, just like you.

Hmm. Got to admit, that does all fit with the latest brain science discoveries about octopuses. So, since you're such a smart creature, could you stick around and help us figure out how brains actually work?

Sure . . . always happy to lend a hand – or eight – to a friend in need.

Great.

OK folks, before we plunge on in, there are two things you should know about brain science, or *neuroscience* – best to use its proper name since you're going to hear a lot about it in this book:

Number 1: Despite centuries of research, nobody truly understands how a human brain works. In fact, nobody's even figured out precisely how the tiny, much simpler, worm brain works.

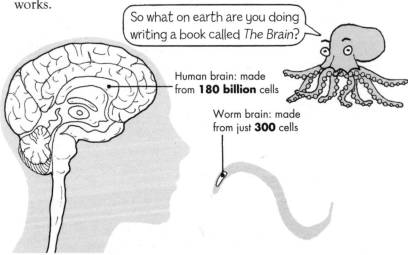

So what on earth are you doing writing a book called *The Brain*?

Human brain: made from **180 billion** cells

Worm brain: made from just **300** cells

It's not that *neuroscientists* are clueless. Far from it; they've made all kinds of amazing discoveries over the years – and more are coming thick and fast today. It's just that the things we 'know' about the brain are still massively outnumbered by all the 'unknowns'. But that's exciting – much of the brain is a huge, mysterious world, just waiting to be explored.

Number 2: Some of the neuroscience discoveries you're going to encounter in this book might sound plain daft – for example:

- Your brain is stuck in the future.
- Your toes don't feel pain.
- A newborn baby has more brain *cells* than you do.
- Everyone might see the colour blue differently.
- Bees can play football.

> And what could be dafter than the idea that human brains are always best at everything!?

Ahem. Sure, we'll definitely be looking at some of our brains' weaknesses, as well as the astonishing powers of other kinds of brain – including yours, Octopus.

But if any of you find yourselves doubting anything you read in this book, that's great! You're thinking like a scientist. Because science isn't really about proving 'facts'. Doubts and mistakes are massively important – as we'll find out when we look at how brain research has progressed over the centuries.

And brain researchers all have one thing in common: intense curiosity. A characteristic it seems you share too. After all, your brain is curious enough to want to understand **itself**. Why else would it have instructed you to pick up this book?

All human brains have a raging hunger for learning and thinking. Each day your brain gorges, non-stop, on memories, feelings and ideas – and, believe it or not, every single one of them leaves a lasting imprint inside your head. And that means – just like faces and fingerprints – every human brain is truly, utterly, wonderfully unique.

> Uniquely weird and unpredictable, you mean!

You say that like it's bad thing, Octopus! And, it's true, our brains can cause us strife when we get anxious, depressed or suffer other *mental-health* challenges (we'll tackle these in Chapter 8). But maybe our brains' quirks and sudden shifts of emotion are part of what makes life worth living?!

OK, if we need to get to grips with the massive, mysterious cauliflowers inside our heads, we've got a lot to cover. First up, we're going to get our hands on an actual brain, to see what it's made up of and how it all fits together. Then we'll get an even closer view of the nitty-gritty details by venturing inside it.

> So, let's get on with it! Catch me if you can!

CHAPTER 1
How the Brain Works
PART 1: GETTING TO KNOW BRIAN'S BRAIN

So, your brain looks a bit like a cauliflower. But what is it made from and how does it work? Anyone willing to let us take out their brain so we can have a proper look?

Great! And don't worry, you'll hardly notice it's gone.

Absolutely. Because, weirdly, the brain itself can't sense pain – you won't even feel us lifting it out.

Well, you're holding Brian's entire world! Be gentle – it's fragile.

The contents of Brian's brain may be unique, but the basic parts work the same way in practically all human brains. Let's see what we've got.

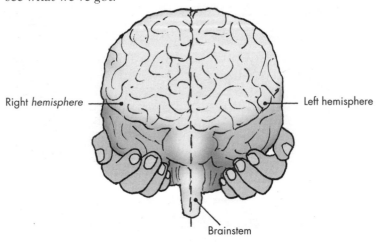

See the line dividing Brian's brain into two halves, or 'hemispheres'? The hemispheres are mirror images of each other, so everything on the left side of the brain has a nearly identical partner on the right side. Brain scientists still don't know why, but the left hemisphere is mainly in charge of the right side of the body and vice versa.

The 'cauliflower' even has a 'stalk'!

Yep - that's the *brainstem*. And Brian wouldn't last a second without it. It takes charge of:

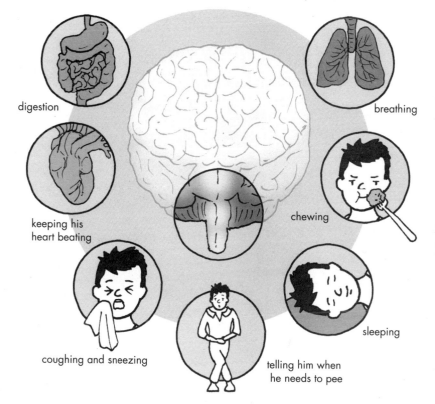

digestion

breathing

keeping his heart beating

chewing

coughing and sneezing

telling him when he needs to pee

sleeping

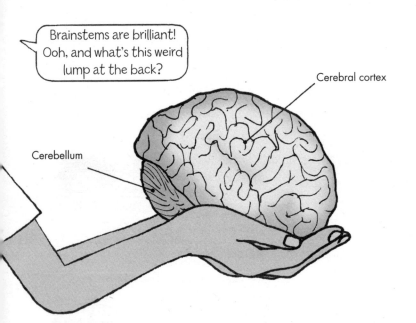

That's the *cerebellum*. It's all about controlling movement and balance. Whenever Brian's walking, dancing, riding a bike, etc. his cerebellum will be working hard to keep him upright and co-ordinated.

Above the cerebellum, the brain's outer layer is called the *cerebral cortex* (or 'cortex' for short). It's so big it only fits inside Brian's skull because it's scrumpled up – hence all those bulges and crevices. If we stretch it out flat, the cortex is as big as a pillow!

It needs to be big because it has huge amounts of work to do. And, more than any other part of the brain, the cortex makes Brian the unique person he is. It takes charge of language, deliberate movements, planning, imagination and personality, as well as stuff that other animals can't do, like algebra and navigating social media. On top of all that, it makes sense of almost everything Brian can see, hear, feel, taste and smell.

There's a heck of a lot going on in the cortex and, like most brain regions, it's always multi-tasking – with lots of different processes happening at once. Some parts of the cortex are mainly concerned with very specific tasks, though.

Let's scrumple Brian's cortex back up and see which bits do what:

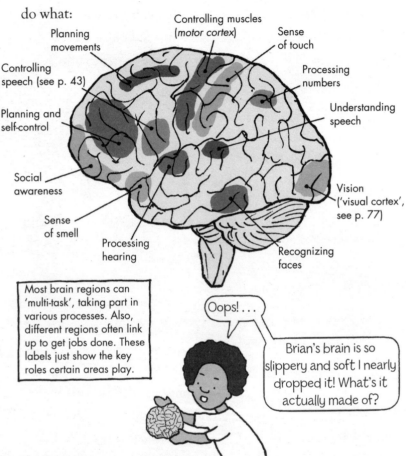

Planning movements

Controlling muscles (*motor cortex*)

Sense of touch

Processing numbers

Controlling speech (see p. 43)

Understanding speech

Planning and self-control

Social awareness

Vision ('visual cortex', see p. 77)

Sense of smell

Processing hearing

Recognizing faces

Most brain regions can 'multi-task', taking part in various processes. Also, different regions often link up to get jobs done. These labels just show the key roles certain areas play.

Oops!...

Brian's brain is so slippery and soft I nearly dropped it! What's it actually made of?

Believe it or not, any brain is mostly water. Even the 'solid' parts aren't all that solid. They're mainly made of fat – oily *molecules* that scientists call *lipids*. All your best ideas and fiercest feelings emerge from as much fat as you'd get in a pack and a half of butter!

And the crucial working parts at the centre of your brain's structure are its microscopic cells. All living things are made of cells.

> So it's the cells in the brain that are made of fat?

Yes, partly. All cells have a kind of 'skin', called a *cell membrane*, which is the bit that's made from lipids. Cells are life's building blocks.* And it takes 180 billion of them to build a brain.

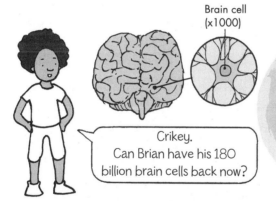

Brain cell (x1000)

> Crikey.
> Can Brian have his 180 billion brain cells back now?

If each one of these 180 billion cells was a penny, this stack would reach two-thirds of the way to the moon

Not yet! There's another bunch of brain parts you need to know about. It's called the *limbic system*, and life would be dull, short and (literally) forgettable without it, because it mainly handles memories and emotions – as well as keeping the body

*Find out more in Explodapedia: *The Cell*.

21

up and running. But the limbic system is buried deep inside Brian's brain. Let's peel back his cortex to take a look:

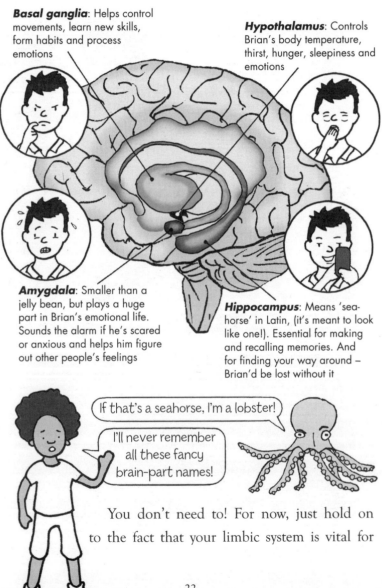

Basal ganglia: Helps control movements, learn new skills, form habits and process emotions

Hypothalamus: Controls Brian's body temperature, thirst, hunger, sleepiness and emotions

Amygdala: Smaller than a jelly bean, but plays a huge part in Brian's emotional life. Sounds the alarm if he's scared or anxious and helps him figure out other people's feelings

Hippocampus: Means 'seahorse' in Latin, (it's meant to look like one!). Essential for making and recalling memories. And for finding your way around – Brian'd be lost without it

If that's a seahorse, I'm a lobster!

I'll never remember all these fancy brain-part names!

You don't need to! For now, just hold on to the fact that your limbic system is vital for memories, emotions and some basic bodily functions.

Right. Brian can have his brain back now . . .

Now Brian can think, feel, see, move, eat, drink, breathe and remember again. His brain – like all our brains – is essential to pretty much every aspect of being human.

It's a bit complicated. Maybe the best way to show you is by taking a (guaranteed, pain-free) trip inside Brian's brain. OK with you Brian?

PART 2: Wiring Up Brian's Giant 'Supercomputer' Brain

Prepare yourself: we're about to shrink down to just one hundredth of a millimetre tall. Then we'll climb into this tiny, remote-controlled transport capsule ready to be fired up Brian's left nostril and into his brain.

Fun! Can I come?

Sure. But it's going to be pretty weird in there – a bit like entering the heart of a massive *supercomputer*. That's roughly what a human brain is: a huge 'warehouse', packed with powerful computers.* The 'warehouse' is the main 'control centre' for your entire *nervous system* – the gigantic 'communications network' that's linked to every single part of your body.

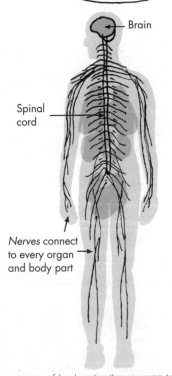

Brain

Spinal cord

Nerves connect to every organ and body part

---

* Your brain isn't literally like a computer – it's way more powerful and creative than any computer invented so far.

24

OK, we're in.

Yep, it's wet and wild in here.

Now, let's take a look at the individual parts that nervous systems are built from – those fat-coated cells we heard about on page 21. All animals' nervous systems are made from two different kinds of cell: *neurons* and *glia*. Let's check out the neurons first.

1. Neurons: The communications experts

Neurons are the 'wires' that ferry information – in the form of *electrical impulses* and chemical signals – around Brian's nervous system.

Doesn't look anything like a wire!

Well, they are long and stringy. Most other cells are roundish blobs. Neurons are expert communicators because of these key parts:

Dendrites: **Receive** signals **from** other cells. Some neurons have **thousands** of dendrites, and communicate with **thousands** of other cells, all at the same time

Cell body: Contains *nucleus* that's packed with *genes* (instructions for building and operating the cell)

Axon: Main 'cable' **sending** signals **towards** other cells

Axons are super skinny and can be extremely long – some can stretch from the spine all the way to the tips of the toes. If those axons were as thick as a piece of spaghetti, they'd be **four kilometres long**!

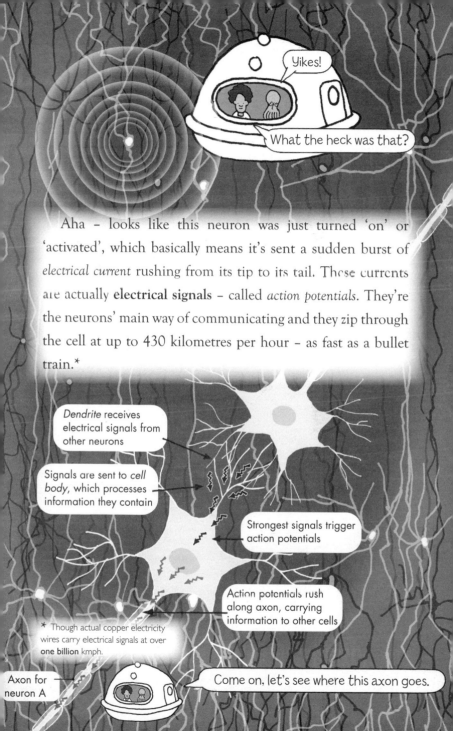

Yikes!

What the heck was that?

Aha – looks like this neuron was just turned 'on' or 'activated', which basically means it's sent a sudden burst of *electrical current* rushing from its tip to its tail. These currents are actually **electrical signals** – called *action potentials*. They're the neurons' main way of communicating and they zip through the cell at up to 430 kilometres per hour – as fast as a bullet train.*

Dendrite receives electrical signals from other neurons

Signals are sent to *cell body*, which processes information they contain

Strongest signals trigger action potentials

Action potentials rush along axon, carrying information to other cells

* Though actual copper electricity wires carry electrical signals at over **one billion** kmph.

Axon for neuron A

Come on, let's see where this axon goes.

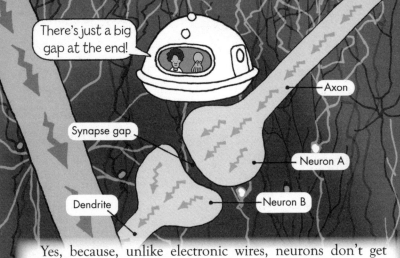

Yes, because, unlike electronic wires, neurons don't get 'plugged in'. For a signal to get from one neuron to the next, it has to hop across that gap, or *synapse*.

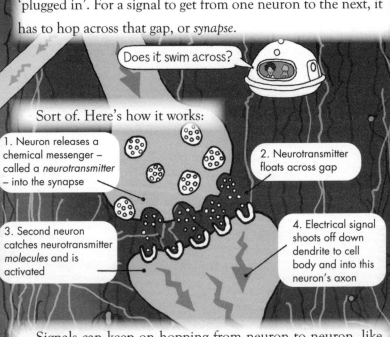

Sort of. Here's how it works:

Signals can keep on hopping from neuron to neuron, like the runners in a relay race.

The human brain uses over 100 different neurotransmitter chemicals and, just as each word in a language has a different meaning, each neurotransmitter can have a different effect on its neuron neighbours, e.g. some turn certain neurons on and some turn certain neurons off.

Just as computers work by processing complicated patterns of ones and zeroes, everything our nervous systems do for us – all thoughts, sensations and bodily instructions – depend on different patterns of neurons that are either on (i.e. sending action potential signals) or off.

Got it. So Brian's brain is basically packed with billions of these on/off neuron wires?

There's other stuff in there as well, including the crucial glia cells.

2. Glia: The workers that keep the brain going

Neuron

Glia

There are so many of them! They're beautiful!

Yeah – this one looks like me!

The glia do all kinds of essential jobs, including:

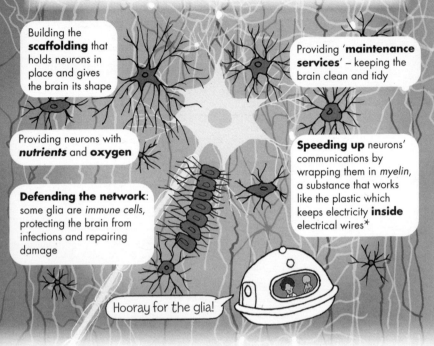

Building the **scaffolding** that holds neurons in place and gives the brain its shape

Providing 'maintenance services' – keeping the brain clean and tidy

Providing neurons with **nutrients** and **oxygen**

Speeding up neurons' communications by wrapping them in *myelin*, a substance that works like the plastic which keeps electricity **inside** electrical wires*

Defending the network: some glia are *immune cells*, protecting the brain from infections and repairing damage

Hooray for the glia!

Yep, neurons couldn't send a single message without them.

Right. Let's shoot back out through Brian's nostril and de-shrink ourselves. Then we can get to grips with how all these brain cells, working together, create the body's communications network.

---

* Myelin is mainly made of lipids – another big reason it takes so much fat to build a brain (see pp. 20-21).

3. 'In the blink of an eye' – Neurons form circuits that do stuff

Neurons and glia weave their magic by forming *neural circuits* – loops of connected neurons that team up to get stuff done.

A *reflex* is a simple type of neural circuit. Reflexes detect specific things that are happening to your body and then **automatically** trigger quick, helpful **actions**.

Suppose Brian is riding his bike when a pesky little fly zips into his eye. There's a type of reflex that makes his eye blink. Here's how it works:

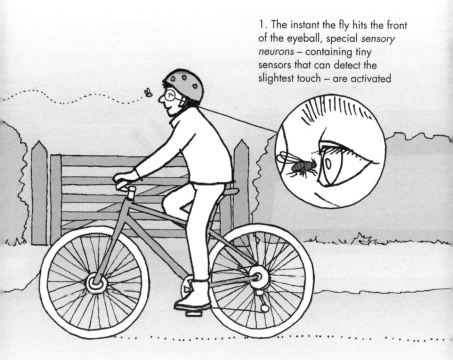

1. The instant the fly hits the front of the eyeball, special *sensory neurons* – containing tiny sensors that can detect the slightest touch – are activated

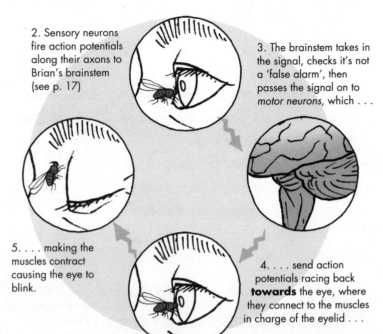

2. Sensory neurons fire action potentials along their axons to Brian's brainstem (see p. 17)

3. The brainstem takes in the signal, checks it's not a 'false alarm', then passes the signal on to *motor neurons*, which . . .

5. . . . making the muscles contract causing the eye to blink.

4. . . . send action potentials racing back **towards** the eye, where they connect to the muscles in charge of the eyelid . . .

All this happens in less than **one tenth of a second,** swishing the fly out of Brian's eye, so he can carry on cycling.

Everything your brain and nervous system do for you is based on neural circuits that gather and process information and use it in the same basic way.

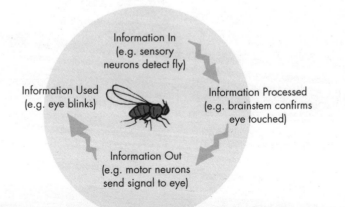

Information In
(e.g. sensory neurons detect fly)

Information Used
(e.g. eye blinks)

Information Processed
(e.g. brainstem confirms eye touched)

Information Out
(e.g. motor neurons send signal to eye)

I thought you said brains were complicated?

Well, think about it this way, Octopus. The entire internet contains around 20 billion connected computers. A full-grown human brain, on the other hand, contains a whopping 86 billion neurons and a similar number of glia. Neuroscientists estimate that those neurons are linked together by an astonishing **600 trillion** synapses.

Sounds like quite a lot of synapses.

Yup. There are more synapses in one brain than there are stars in one thousand galaxies! Together, they form a truly gargantuan number of neural circuits, some of which are brain-bogglingly complicated.

Neuroscientists deserve serious respect for daring to try and figure out how those buzzing circuits inside your head transform a jelly-like, cauliflower-shaped lump of fatty matter into an awesomely intricate brain that lets you think, feel and do!

So, let's get to know some of the trailblazing scientists and thinkers who've tackled this huge challenge over the years?

I'm up for that!

And I need a lie down. Have fun, guys!

CHAPTER 2
Brain Exploration
THE PIONEERS WHO PEERED INSIDE OUR SKULLS

For thousands of years, most people assumed the brain was a pointless lump of pinkish-grey jelly. Some ancient Egyptians, for example, thought its main purpose was to make snot. So, when they mummified a body, they yanked the brain out through the nose and chucked it in the bin.

Hey, what about my jar?!

They reckoned the heart was in charge of thoughts and feelings. Thousands of years later, hints of that belief still linger, like when someone says, 'you know it in your heart'.

You won't meet a single neuroscientist who genuinely thinks the heart can 'know' anything, however. Today, they're convinced thoughts and feelings happen in the brain.

But how were ideas about the brain propelled from pointless snot factory to all-powerful conjuror of thoughts, feelings and actions? Let's meet some of history's most intrepid scientists to find out.

It's All about the Brain (ancient Greece, 400 BCE)

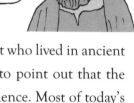

From the brain and from the brain only arise our pleasures, joys, laughter and jests, as well as our sorrows, pains, griefs and tears!

This is Hippocrates, a doctor and scientist who lived in ancient Greece. He was one of the first people to point out that the brain was at the core of all human experience. Most of today's neuroscientists would pretty much agree with his statement.

But how did Hippocrates figure it out?

By careful observation of his patients' symptoms, Hippocrates showed that different brain injuries can cause all kinds of specific problems, including blackouts, fits, paralysis and changes of mood and personality. When Hippocrates was treating a wound on the **right** side of a soldier's brain, for example, he noticed that the injury affected actions on the **left** side of the body. He concluded, correctly – nearly 2,500 years ago! – that each brain hemisphere controls the **opposite** side of the body (see p. 17).

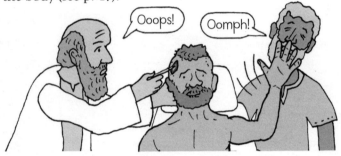

Five centuries later, a Roman doctor called Galen tried to figure out how the brain worked.

IS THE BRAIN A PUMP? (ROME, C. 170 CE)

Galen thought the best way to understand brains was by dissecting them. That's still a totally valid scientific technique today, it's just that Galen sometimes used **live** animals . . .

In one clever but cruel experiment, Galen sliced through a *nerve* connecting a pig's brain to its throat. The pig seemed fine – except it stopped squealing. Not nice for the pig, but it gave Galen solid scientific evidence that Hippocrates was right: the brain controlled the whole body, including the *vocal cords*.

Galen's explanation for how the brain actually functioned was less scientific. He thought of it as a kind of pump that pushes fluids, called humours (blood, phlegm, yellow bile and black bile) around the body, through nerve 'pipes'.

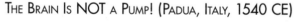

They weren't. According to Galen, the four humours controlled our emotions, personalities, movements and even our body shape.

This rather dubious idea stuck around for 1,400 more years!

THE BRAIN IS NOT A PUMP! (PADUA, ITALY, 1540 CE)

Then Andreas Vesalius pointed out that Galen's life's work, which doctors had accepted as fact for centuries, was full of mistakes.

Dissecting dead human bodies had been illegal in Galen's day, which is why he'd based his observations on **other** animals' bodies.

Vesalius, however, carved up actual human corpses and described every minute detail of all human organs, including the brain – and he declared that Galen's 'brain pump' idea was wrong!

This thing can't pump! And where are the pipes?

He didn't really have a better idea, though.

Not looking so smug now, are you?

Hmmph.

Machines with Minds? (Paris, France, 1640 CE)

A century later, René Descartes, philosopher, mathematician, scientist and all-round mastermind, decided to solve the brain mystery once and for all, without dissecting a single brain.

He'd become obsessed with incredibly complicated clockwork 'automata' (basically, mechanical robots) that people were building at the time . . .

Descartes suggested that all animals, including humans, must work in a similar way. In other words, he thought our bodies were extremely complex **machines**, with the brain as the main control centre – which is not so far from what many scientists think today.

But Descartes reckoned no machine, however complex, could ever truly 'think' or 'feel'. He suggested some kind of invisible spirit – he called it a 'rational soul'– must live inside the brain, yet be separate from it. This part of Descartes's thinking is the basis of a theory philosophers call dualism.

He didn't know. But Descartes's dualism still influences the way we think about our brains. His rational soul is what we might call the *mind* – i.e. the brain processes that think, feel and remember, making you who you are.

Unlike Descartes, most scientists today would say the mind definitely **is** created by the complex workings **inside** the brain and body. But they'd probably agree that the mind itself isn't really a solid, physical thing that they can measure or touch.

Some might even agree with Descartes's bold claim . . .

It means, 'I think, therefore I am,' in Latin (the language Descartes wrote in). It was Descartes' way of pointing out that our minds are probably **the only things** in the entire world we can know for certain actually exist. He proved this by asking us to imagine that **everything** we're experiencing is actually imaginary – as if we are each living in a massive *virtual reality* simulator controlled by an 'evil demon'.

It's enough to make you doubt everything. Everything, Descartes argued, apart from the fact that you are doubting your experiences. And doubting is a kind of 'thinking', so there must be a 'thinker'. In other words, your **mind** must exist!

But Descartes probably never imagined that our complex, machine-like brains run on electricity . . .

AN ELECTRIFYING SHOW (LONDON, UK, 1803)

When Italian scientist Giovanni Aldini put on a live show for an audience in London, his co-star, Matthew Foster, just happened to be dead. He'd recently been hanged for murder. Aldini connected a massive battery to Foster's face, turned it on and . . .VOOOM.*

* Aldini's shows may have inspired Mary Shelley to write *Frankenstein*, in 1816, about an eccentric doctor who builds a monster from human body parts and brings it to life using electricity.

The jolt of electricity made Foster's dead face grimace and his eye open. Some thought the electricity was bringing the dead man back to life!

> That's just sick. You can't call it science, surely?

Actually, other scientists were doing similar 'experiments' at the time, and the logical conclusion they reached was that brains control bodies by making their own electricity.

> So now you're telling me my brain's an electricity generator?!

Not exactly. But each neuron does make a tiny amount of electricity. Your whole brain uses about as much electricity as it takes to power an *LED* lightbulb. But it does amazing things with that tiny amount, as we'll soon see.

Brain Regions Speak Up (Paris, France, 1860)

Neurologist Paul Broca was particularly intrigued by a patient called Louis Victor Leborgne, who'd suffered a *stroke* that had left him able to say only one word:

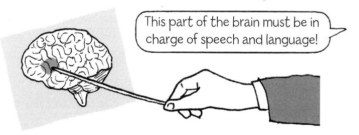

Leborgne's overall *intelligence* and awareness seemed fine, so Broca reasoned that one specific part of his patient's brain must have stopped working.

About a year later, Leborgne died, and Broca was able to examine his brain. The doctor found a small damaged area . . .

He was right! It's still called 'Broca's area' today.

Broca's work revealed two key lessons:

• Some brain regions focus on particular jobs (see Chapter 1).

• Studying the way brain damage affects specific functions – e.g. speech, memory or emotion – can reveal what the damaged parts normally do.

But the brain regions Broca studied each contained millions of neurons. How did they work together to create amazing things like speech? To find out, we've got to zoom right in . . .

DRAWING OUT THE BRAIN'S LIVING WIRES (BARCELONA, SPAIN, 1890)

Spanish scientist Santiago Ramón y Cajal knew the brain worked by making, sending and receiving electrical signals (see p. 27), and he was determined to find out how its 'wires' actually functioned.

So, he cut wafer-thin slices of brain tissue and then treated them with a special stain that made single brain cells appear dark brown on a white background.

He couldn't believe the beautiful branching, tree-like structures that appeared before his eyes. Luckily, Ramón y Cajal

was also a brilliant artist, and he sketched hundreds of neurons and glia cells for all to see.

But these weren't just pretty pictures. Ramón y Cajal showed that:

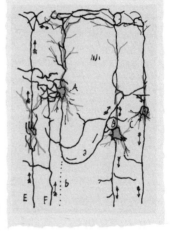

• neurons connect to each other via synapses (see p. 28)

• neurons are very particular about which other cells they make synapses with

• signals only flow one way, from the dendrites to the cell body and **down** the axon (see p. 26).

Ramón y Cajal's groundbreaking discoveries, particularly his insights into synapses, were a springboard for future neuroscientists. It's these connections between brain cells that form the neural circuits that are behind almost everything our brains think and do (see p. 31).

Even so, it took another 30 years to figure out how signals cross the synapse 'gaps' . . .

45

GETTING TO THE HEART OF THE SYNAPSE (GRAZ, AUSTRIA, 1921)

Austrian scientist Otto Loewi knew a frog's heartbeat was controlled by the *vagus nerve*, which is connected by a synapse to the heart muscle. When he activated the vagus nerve with electricity, the frog's heart slowed down: a signal from the nerve must have jumped **across** the synapse into the muscle. But how?

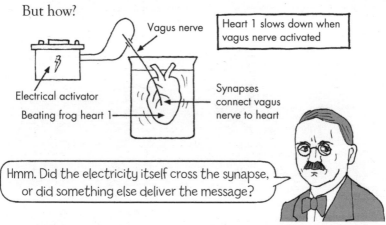

Vagus nerve

Heart 1 slows down when vagus nerve activated

Electrical activator

Beating frog heart 1

Synapses connect vagus nerve to heart

Hmm. Did the electricity itself cross the synapse, or did something else deliver the message?

To find out, Loewi took fluid from around the activated synapses and added it to another frog's heart. Even though its vagus nerve had been removed, the second heart slowed down.

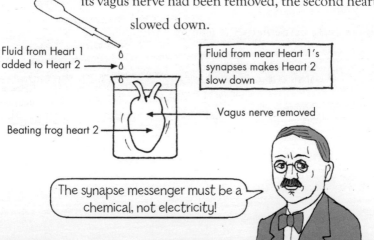

Fluid from Heart 1 added to Heart 2

Fluid from near Heart 1's synapses makes Heart 2 slow down

Vagus nerve removed

Beating frog heart 2

The synapse messenger must be a chemical, not electricity!

Loewi had discovered the very first neurotransmitter (see pp. 8-28). But a big question remained: How do neurons generate the electrical signals that trigger those neurotransmitters?

SQUID SHOW GREAT (ACTION) POTENTIAL (PLYMOUTH, UK, 1948)

When Alan Hodgkin and Andrew Huxley moved their entire lab to the seaside, they weren't interested in surfing, they just wanted fresh squid.

They contain neurons with 'giant axons' up to **one millimetre wide** which, unlike almost all other neurons, are visible to the naked eye.* They are so big, Hodgkin and Huxley could easily jab thin *electrodes* into them, allowing the scientists to measure, and alter, the pulses of electric current that flowed through the neurons.

They carried out hundreds of delicate experiments.

*The axons in your nervous system are on average just 10 microns across – 100 times narrower than squids' giant axons.

Sadly, they did. But those squid helped the researchers achieve a massive breakthrough: they figured out exactly how action potentials – the pulses of electricity triggered when neurons are activated (see p. 27) – work. It all comes down to the way neurons control the flow of electrically charged *ions* into and out of a cell, across their cell membranes (see p. 21).

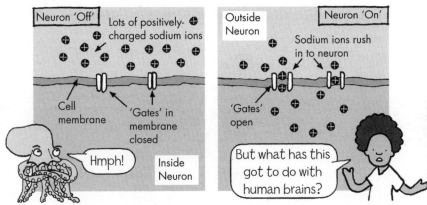

Everything. The way neurons work is the **same** in all animals. Many of the biggest discoveries in neuroscience, from Galen and his pigs (see pp. 36-37) onwards, have come from studying non-human animals.

Studying them? Don't you mean **torturing** them?!

Nowadays, most neuroscientists try to avoid using animals in experiments, and when there's no other good option, they do their best to make sure the animals don't suffer too much.

It's hard to know if sea slugs suffer, but we do know that these critters helped a brilliant young neuroscientist called Eric Kandel to understand precisely how brains learn things and store memories.

MEMORIES LEAVE THEIR MARKS (NEW YORK, USA, 1970)

Ever been left with the jitters ages after watching a scary movie? A neuroscientist would say you've been *sensitized* to creepy *stimuli*, like creaking floorboards or howling dogs. Sensitization is a kind of learning – the movie taught your nervous system to fear the worst.

Kandel invented a technique that worked in a similar way to sensitize sea slugs, so they overreacted to the slightest touch.

He then examined the sea slugs' nervous systems to see how sensitization learning had altered their neural circuits. It turned out that the synapses in the slugs' reflex circuits (see p. 31) were 'strengthened', meaning they were more easily activated by neurotransmitters (see p. 28).

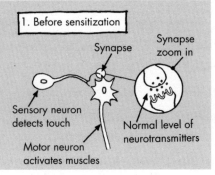

1. Before sensitization

Synapse

Synapse zoom in

Sensory neuron detects touch

Normal level of neurotransmitters

Motor neuron activates muscles

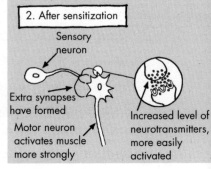

2. After sensitization

Sensory neuron

Extra synapses have formed

Motor neuron activates muscle more strongly

Increased level of neurotransmitters, more easily activated

However, this turned out to be about so much more than just sensitization learning. Thanks to Kandel's breakthrough, we now know that practically everything we learn – from maths to table manners – and absolutely all our memories, are based on making, breaking or adjusting the synapses that hook up neural circuits. So, if you remember anything about this story of sea slugs, that's because reading about it has **changed your brain**!

Of course, our brains are a bit more complicated than sea slugs' brains . . .

Here we go again . . .

. . . and we can't keep opening people's skulls to take a look at what's going on inside . . .

No you cannot!

Neuroscientists really needed a machine that could look straight **through** the skull and see living brains at work.

Seeing Brains without Sawing through Skulls
(Nottingham, UK, 1978)

Physicist Peter Mansfield knew traditional X-ray pictures were useless for examining brains, since they only showed solid body parts, like bones. To solve this problem he dreamed up an extraordinary technique called *magnetic resonance imaging* (MRI).

Bodies, like everything else in the world, are made from tiny building blocks called *atoms*. Mansfield's MRI scanner used incredibly strong magnets and radio waves to visualize soft body parts, by detecting which atoms they're made from and how densely packed they are.

But once the first MRI scanner was finally built, Mansfield's colleagues were worried its huge *magnetic field* was unsafe.

So Mansfield climbed inside the gigantic, noisy new contraption himself. It worked – and no harm was done!

Nowadays, MRI scanning is one of the best ways of seeing what's going on inside our brains. It can't zoom in on individual neurons, but it can show which brain regions are most active when people are thinking in particular ways.

Mansfield's scanner couldn't read his thoughts, but one cutting-edge invention may be getting close to doing just that . . .

MIND READING (MELBOURNE, AUSTRALIA, 2021)

In 2021, Dr Thomas Oxley and his team inserted a tiny device called a *brain-computer interface* (BCI) into a patient's neck. Then they pushed it in until it reached a blood vessel next to the part of the cerebral cortex that controls movement (the 'motor cortex'). The patient, Philip O'Keefe, has 'amyotrophic lateral sclerosis' (ALS), a disorder sometimes known as motor neurone disease, which affects his nerve cells, preventing him from moving his hands.

After a bit of training, O'Keefe was able to operate his computer . . . by **mind control**.

I just **think** about where I want to click, and I can email, bank, shop, and now message the world via Twitter.*

* Now called 'X'.

The BCI is literally reading his mind. Its sensors listen to neurons in O'Keefe's motor cortex and send the signals, that would usually go to his hands, to a computer. The system recognizes patterns of neuron activations that let it understand exactly what he wants it to do.

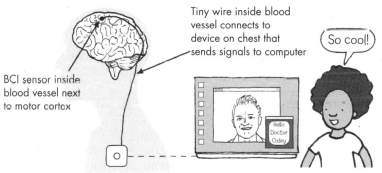

This may just be the beginning, but BCI technology is improving fast. Before long, neuroscientists might be able to decipher all kinds of more complicated thoughts and feelings. It could eventually lead to BCIs that cure *mental health conditions* or even give us totally new mental abilities that we can scarcely imagine (more on this later).

But all that's still way off in the future, which, as we're about to see, is your brain's very favourite topic.

53

CHAPTER 3
What Brains Are 'For'
THE LIVING MACHINES THAT PREDICT THE FUTURE

What would you say your brain is for?

Here's the answer in a nutshell: brains are for keeping us alive. And the main way they achieve that is by **predicting** what's going to **happen next** and then **doing something** about it.

Crikey - my brain did not predict you were going to say **that**!

Ha! No brain can predict the future perfectly, especially in a world like ours where things are always changing. Nevertheless, lots of neuroscientists think brains basically exist to help the living bodies they're part of stay alive and well for as long as possible. A huge part of that is trying to make sure they avoid nasty surprises – like getting attacked by predators, or running out of food.

Smart Cells

That's true for humans, with our super complex brains, and for simpler organisms that don't even know what a nasty surprise is – like these *bacteria*. All cells need systems for making predictions and adjusting their behaviour in a constant effort to survive.

x5000

How can that tiny little blob predict anything?

No need to be rude, Octopus. It can, and here's how:

Bacterial cell

1. 'Swims', by rotating whip-like tail called a flagellum

2. Detects sugar molecules using sensors embedded in cell membrane

3. Feeds by absorbing sugars from its surroundings

Problem:

1. Bacteria cell has 'eaten' most of the nearby sugar

3. When it moves in this direction, sensors detect traces of sugar

2. Cell starts to swim, zigzagging around in random directions

Cell makes prediction: There's probably more sugar that way

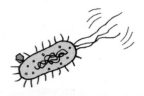

Cell takes action: Swims off towards traces of sugar, 'hoping' to find its next meal. Like a dog following a scent trail, bacteria cell steers towards the place where most sugar is available

Problem solved: Bacteria cell carries on eating

OK. Maybe it isn't as stupid as I thought.

Exactly. But all brains work in the same basic way. They gather information, decide whether they're in a good or bad situation, make predictions, then leap into action.

That's actually pretty much how some scientists define 'intelligence': it basically means 'using information to solve **problems**'.

So, all individual living cells, including all those that make up your body, have some basic form of intelligence. But when lots of neuron and glia cells link together to form a nervous system, they get much, much better at making predictions, and using them to solve more and more complex problems.

Just think about what your brain can do for you when you get hungry.

Tummy rumbles. **Problem:**

Chances are your brain (or maybe the brain of someone who cares about you!) has already predicted this will happen. With luck, there's already something tasty in the fridge.

And your brain can almost certainly predict what it needs to do in order to get hold of that food, cook it (if necessary) and shovel it into your mouth!

Problem Solved:

Remind you of the bacteria cell we just saw? Thanks to our brilliant brains, we humans can make more detailed guesses about what's heading our way and hatch more sophisticated plans.

Then, when individual people combine their brainpower, they can see **much** further into the future and devise even more complicated ways of avoiding unpleasant surprises, like hunger. Just think of all the different people who contributed to your last meal:

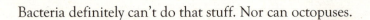

Bacteria definitely can't do that stuff. Nor can octopuses.

Yeah, yeah, humans are amazing, we know . . .

The thing is, our super-complicated brains didn't appear out of nowhere. Like all brains, they *evolved* gradually, over millions of years, through a long series of different species – our *ancestors* – who each had ways of predicting events and controlling their bodies.*

We can actually see the stages of evolution our ancestors' brains went through by looking at certain species that are still alive today.

*Find out how evolution works in *Explodapedia: Evolution*.

How to Evolve a Human Brain in Six Stages

Stage 1. Simple Neuron-like Cells (First evolved ˜700 million years ago)

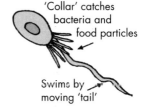

'Collar' catches bacteria and food particles

Swims by moving 'tail'

Living examples: Choanoflagellates (say *ko-anno-fladge-ell-ates*) – microscopic predators.

Key abilities: Can move quickly and catch bacteria by sending electrical signals, a bit like action potentials (see p. 47), shooting from one end of their single-celled bodies to the other.

Future forecasting ability: Very low. Mainly live in the moment, reacting as things happen.

Stage 2. *Nerve Nets* and *Nerve Cords* (First evolved ˜550 million years ago)

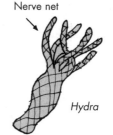

Nerve net

Hydra

Living examples: Jellyfish, sea anemones, starfish, earthworms – animals with complex, *multicellular* bodies.

Key abilities: Can move and feed by using their nerve nets and nerve cords to co-ordinate their muscles and other organs. Nervous systems also allow better sense organs, like eyes, to develop.

Future forecasting ability: Low. Having senses allows slightly more detailed predictions, e.g. working out when a hungry predator is approaching.

Stage 3. Brain-like *'Ganglia'* (First evolved ~500 million years ago)

Living examples: Spiders, lobsters, flies, ants – their heads contain large clusters of neurons and glia called 'ganglia'.

Ganglia

Key abilities: Extra processing power means they can store simple 'maps' in their heads and use them alongside information from smells, tastes, colours and shapes.

Future forecasting ability: Medium to high. Improved ability to predict other creatures' behaviour, find food and stay out of harm's way.

Stage 4. Actual Brains (First evolved ~450 million years ago)

Living examples: Fish, frogs, birds, reptiles – all have a brainstem, cerebellum and limbic system, but the cerebral cortex (see Chapter 1) is generally small or absent.

Cerebellum   Cortex
Lymbic system

Key abilities: Improved understanding of the world and greater flexibility than most Stage 3 animals, e.g. can learn new behaviour patterns when a new source of food appears.

Future forecasting ability: High to very high. For example, when many birds and fish migrate they blend detailed memories with cues from the seasons, the moon and the Earth's magnetic field to predict the best route to their destination, often thousands of kilometres away.

Stage 5. Mammalian Cerebral Cortex (First evolved ~200 million years ago)
Living examples: Rats, dogs, monkeys – increased thinking power, thanks to enlarged cortex containing many more neurons.

Cortex

Cerebellum

Lymbic system

Key abilities: Even greater scope for communicating and learning. Wider range of emotions, e.g. joy, fear, anger and sadness, make forming social groups possible (e.g. wolves hunt as a pack).

Future forecasting ability: Very high. Can predict various future events and choose between different possible actions.

Stage 6. Enormous, Scrunched-up Cerebral Cortex (First evolved ~2 million years ago)

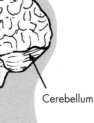

The human cortex is folded many times to fit inside the skull

Lymbic system hidden beneath cortex

Cerebellum

Example organisms: Humans – massively expanded cortex, particularly front parts in charge of planning, making decisions and social interactions.

Key abilities: Using complex language and creating technology, art and science.

Future forecasting ability: Extremely high. They can devise intricate plans and imagine possible futures that have never existed before.

> Ahem. So my exceptionally brilliant brain comes next, at Stage 7, right?

Hmmm, not exactly. But the six stages of brain evolution above aren't necessarily step-by-step 'improvements'.

For example, even though birds (Step 4) haven't evolved a mammal-like cortex (Step 5), lots of them can do incredibly smart things. Clark's nutcrackers can bury seeds in thousands of different locations and remember where they put all of them when they're hungry, many months later.

And even though some animals' brains seem to have barely changed for millions of years, they're not outdated. The creatures you've just met may resemble some of your ancestors, but the reason they're all still around today is that evolution kitted them out with nervous systems that are brilliant at keeping them alive.

INTELLIGENT ALIENS?

Your brain, however, Octopus, is basically unique. The last time we shared an ancestor with you was at least 550 million years ago (Stage 2). Since then, we've been on very different evolutionary journeys.

Nowadays, a typical octopus has about 500 million neurons (about the same as a cat).

A third of its neurons are in its doughnut-shaped central brain

Each arm contains its own 'mini brain'

Oi!

Don't blame me, my arms have minds of their own!

Octopus brains are very, very different from ours, but there's no doubting their fiendish intelligence. They have excellent memories and can solve tricky puzzles – like escaping from the **inside** of a jam jar with its lid screwed on tight. And they're excellent communicators – scientists think one of the main ways they do this is by sending dazzling colours and patterns rippling across their skin. Researchers just haven't got a clue what octopuses are saying!

There's even some evidence that octopuses can 'see' with their skin. It's almost impossible for us to imagine how that might feel, which is one of the reasons octopus expert Peter Godfrey-Smith said an encounter with an octopus *'is the closest we will come to meeting an intelligent alien.'*

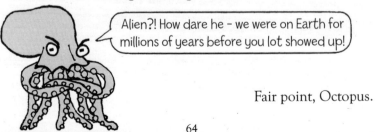

Fair point, Octopus.

And many scientists suspect that, just like us humans, individual octopuses do indeed have rich and varied emotional lives and distinct personalities. You wouldn't want to get on the wrong side of a grumpy octopus. For example: one scientist in New Zealand got squirted with two litres of water whenever she dared walk into the lab an octopus lived in!

Well, an octopus never forgets! She must have done something to deserve it.

Hey! Ugh!!

So . . . is a good memory really a sign of high intelligence?

Partly, yes. Intelligence is about using information to solve problems, remember (see p. 56)? So, learning from the **past** is usually the best way of forecasting the future and therefore avoiding unpleasant surprises and staying alive.

Memory: Looking Back so We Can Look Forward

Memories hold traces of the people you've met, the conversations you've shared, the things you've read and watched, the skills you've mastered and much, much more. They have an outsized role in making you the person you are **today**. And since they're right at the centre of your brain's forecasting abilities, they'll affect pretty much everything you do in the **future**.

Neuroscientists still have a lot to learn about memories – they don't actually know where most of them are stored – but they can tell us the main **kinds** of memory our brains use:

MEMORY

SHORT-TERM MEMORIES
usually last for just seconds or minutes

Sensory memories

provide your brain with a constant stream of snapshots of the things you're seeing, hearing and feeling. Like a camera flash repeatedly lighting up a dark forest, they give your brain quick glimpses of everything around you, before fading away or being stored as longer-lasting memories

Working memory

is your brain's notepad and pencil. It helps you remember shopping lists, do sums, sketch out plans and play with ideas. But each 'notebook page' only has space for 3–5 'chunks' of information at a time, most of which won't become long-term memories

Woof Woof

TO DO
• WALK DOG
• STROKE CAT
• HOMEWORK
• CALL NANA

Implicit memories –

like the computer code that controls a robot – let you do things or react to situations **automatically**, without having to think, e.g. riding a bike or reading a book (though you need to master these skills first)

LONG-TERM MEMORIES

can last from minutes or hours to entire lifetimes

Explicit memories

can be stored and retrieved (hopefully!) on purpose. They come in two main kinds:

Semantic memories –

your brain's own 'Wikipedia'– hold the facts you know, including, perhaps, everything you've absorbed from the first three chapters of this book

Episodic memories –

like a 'movie of your life' inside your head – contain events, places you've been and things that have happened to you, e.g. your first day at school

Now I've got even more names to remember!

Well, some of them have probably found their

way into your *semantic memory* already! Don't worry about all the scientific names, though – not all neuroscientists agree on separating memories into these particular categories. But they do all agree that our memory systems **don't** work like computer hard drives. You can't save memories and then call them up later, completely unchanged. When our brains store *episodic memories*, like your first day at school, for example, they split them up into lots of different parts: faces, places, feelings, words, facts and actions, etc. – and store those parts in circuits scattered throughout the brain.

Recalling that memory is like solving a puzzle: your brain tries to find and slot together as many relevant pieces as it can. But those memories are never perfectly accurate. And that can cause problems if you're a crime-scene witness, for example.

This more flexible way of remembering is actually really helpful. It means you're always updating your memories, by blending your understanding of the past with everything else that you know right now. And that definitely helps your brain complete its main mission in life of . . .

keeping us alive by predicting what's going to happen next and then doing something about it. That bit's lodged in my *explicit memory!*

And my brain's predicting it's time for a new chapter.

Bingo.

CHAPTER 4
All an Illusion
WHY NOTHING YOU EXPERIENCE IS REALLY REAL

Warning: you might find this next statement a bit shocking:
None of the things you experience in your life - from seeing
the vivid colours of a rainbow, to smelling hot buttered toast,
to the warmth of a hug from your best friend - are truly **real**.

According to the latest neuroscience research, they're all
illusions conjured up by your brain and body
to help keep you alive and well.

> Owwwwww! My brain is NOT making this up!

OK, yes, your pain definitely seems
pretty real right now, but those pain
sensations are mainly in your brain.

I don't believe you!

It's absolutely true! Your toe can't feel the damage caused by the rock you stubbed it on. If it wasn't connected to your brain, you wouldn't have noticed a thing.

How come my brain knows what's going on?

Well, that's the really weird bit: it doesn't. Your brain lives its entire life tucked away inside the solid, dark box of your skull; it has no way of knowing what 'reality' actually is, because it hasn't ever seen, heard, felt, smelled or tasted the outside world.

So, your brain's entire view of the world is based on the signals that flow into it from the rest of your body – from your sense organs in particular. Your eyes, ears, nose, tongue and skin communicate with your brain via neurons.

Your brain then has to turn all the electrical blips and chemical pulses (e.g. neurotransmitters, see pp. 28-29) coming from those neurons into the rich and powerful sensations you feel each day. Considering that's all it's got to work with, it does a pretty amazing job, doesn't it?!

But my toe still hurts!

Well, look what happened when you kicked that rock:

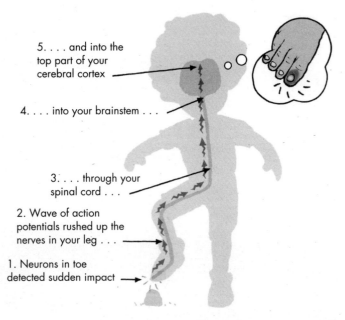

5. . . . and into the top part of your cerebral cortex ———

4. . . . into your brainstem . . .

3. . . . through your spinal cord . . .

2. Wave of action potentials rushed up the nerves in your leg . . . ———

1. Neurons in toe detected sudden impact ———

Then your brain, not your toe, translated those nerve signals into a feeling of blazing agony.

What's more, the way your brain reacts to signals from your senses has a big influence on what you actually feel.

In a 2009 experiment, for example, researchers told volunteers to plunge their hands into painfully cold iced water, instructing half of them to swear out loud when their hand started hurting and the rest to keep schtum. The people who swore said they felt less pain! It turns out that swearing can actually trigger natural 'painkiller' chemicals to be produced inside the brain. That certainly doesn't mean pain is 'made up', but it does show that it has a lot to do with what's going on inside your head.

The same goes for **all** sensory experiences, or *perceptions*. So let's make sense of this strange-sounding idea by taking a closer look at how most brains figure out what their eyes are seeing.

SEEING IS BELIEVING. OR IS IT?

You might imagine that your eyeballs work like little 'webcams', recording images of your surroundings that are then live-streamed straight into your brain.

But if your eyes really did work like that, your world would look very, very different. Much of what you'd see would be in black and white, there'd be huge blank areas in the middle and the 'camera' would be constantly jerking, making you feel sea sick!

To see why, it helps to know a bit more about how the human eye is built:

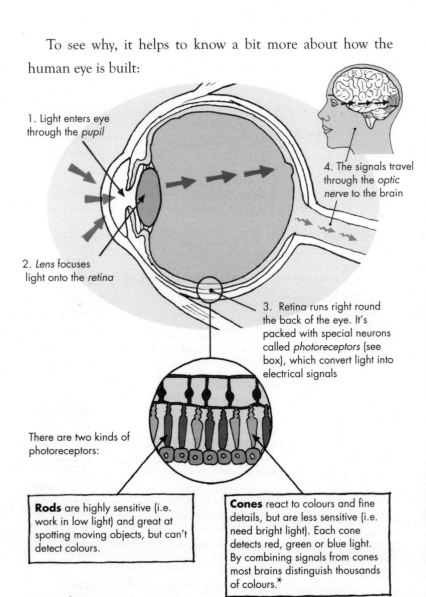

1. Light enters eye through the *pupil*

4. The signals travel through the *optic nerve* to the brain

2. *Lens* focuses light onto the *retina*

3. Retina runs right round the back of the eye. It's packed with special neurons called *photoreceptors* (see box), which convert light into electrical signals

There are two kinds of photoreceptors:

Rods are highly sensitive (i.e. work in low light) and great at spotting moving objects, but can't detect colours.

Cones react to colours and fine details, but are less sensitive (i.e. need bright light). Each cone detects red, green or blue light. By combining signals from cones most brains distinguish thousands of colours.*

*People without all three types of cone see fewer colours, which is sometimes called colour blindness. Some people have four kinds of cone, meaning they can see up to 100 times more colours.

By imagining we're looking through the pupil, to see the retina head-on, we can see how the retina detects patterns of light.

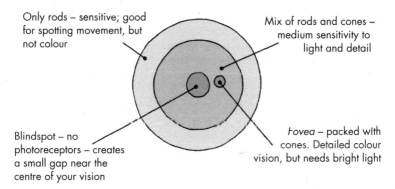

Only rods – sensitive; good for spotting movement, but not colour

Mix of rods and cones – medium sensitivity to light and detail

Blindspot – no photoreceptors – creates a small gap near the centre of your vision

Fovea – packed with cones. Detailed colour vision, but needs bright light

The tiny fovea is the only part that gives you high-resolution vision needed for things like reading this book. It's a bit like you're peering at the world through a narrow tube the width of a pencil, while everything else around it is kind of blurry and dim!

But I can see **everything** perfectly! And the colours are amazing!

That's because your brain doesn't show you **exactly** what your eyeball is detecting. Far from it. To get a complete view of your surroundings, for example, your eyes are constantly moving, so your foveas can take in the most important details. Your eyes are jumping around this page right now. But you're not remotely aware of all that movement.

Your brain does an absolutely incredible job of making sense of all the jerky, gappy, blurry images captured by your eyes. It picks out shapes, colours, moving objects, patterns and faces, for example. This is called 'bottom-up' processing. But, as we're about to see, it's only part of the reason why it feels as if your imperfect eyes give you such a smooth, seamless, full-colour, high-resolution view of your surroundings.

Your Brain Is Making It All Up as It Goes Along

In fact – and here's the really weird bit – after going to a lot of effort to create those bottom-up impressions of the world, your brain quite often simply ignores them!

How come?

It's because brains are all about predicting the future (see Chapter 3). Bottom-up processing takes time – about 0.1 to 0.25 seconds, to be precise – and during that time your brain has already cooked up a 'top-down' prediction of what you're about to see. It bases these predictions on three main 'ingredients':

1. **Short-term sensory memories** (see p. 66) showing what your world was like in the recent past. It's usually fair to assume things don't change massively from one moment to the next.

2. **Long-term memories** of situations similar to some you've encountered in the past. These let you recognize faces, objects and places, for example. So, if you spot a familiar face in a

crowd, your brain can use your memory of that person to fill in any gaps.

3. **Information about what's happening** in your body right now. If you're in a stressful situation, an exam, say, and your heart is racing and hands clammy, your brain will use that extra information to form your top-down prediction about what's going on. Hopefully it will help you react appropriately, e.g. focusing on the questions, rather than what's happening outside the window.

Once your brain has come up with its 'best guess' about what you're about to 'see', it compares that to the stream of information flowing in from the eyeballs. If the top-down predictions and bottom-up data match, the neurons in your visual cortex (see p. 85) will **already** be firing in the correct patterns. And many neuroscientists think your brain basically ditches the bottom-up information. Which is why, most of the time, what you're 'seeing' is a prediction of the near future that's entirely conjured up **inside your brain**.

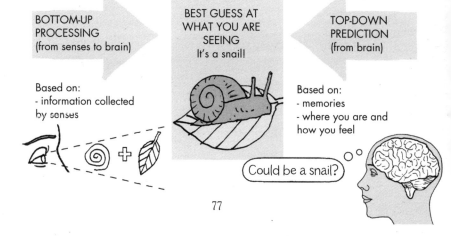

BOTTOM-UP PROCESSING (from senses to brain)

BEST GUESS AT WHAT YOU ARE SEEING It's a snail!

TOP-DOWN PREDICTION (from brain)

Based on:
- information collected by senses

Based on:
- memories
- where you are and how you feel

Could be a snail?

It gets weirder. Your brain's top-down predictions are often less about showing you what's actually in front of you, and more about generating a version of reality that **it thinks** you will find most **useful**.

Luckily for you, your brain's pretty good at interpreting this. Which is why:

Msot plepoe cna slitl raed tihs sntencee evne toghuh teh ltetres aer lal jmubeld up!

It's also why it's so irritating when an author writes a sentence without a last   .

The reason you can still make sense of these two sentences is because your top-down prediction can draw on all its memories – acquired over many years of learning to read, write and talk – of how words and language 'should' work.

GLITCHES IN THE SYSTEM

Your brain's top-down predictions are both accurate and helpful – most of the time. They're what stop you getting confused and wandering out into busy roads or kissing random strangers who look a bit like loved ones, for example. But it doesn't always get things right.

Ever been convinced you can feel a buzzing inside your pocket only to discover there's nobody calling your phone?

That's your brain believing one of its own predictions when it really shouldn't. Perhaps a muscle twitched randomly or your clothing brushed against your skin, setting off a false alarm. And, if you were hoping someone would call, your brain would probably be more likely to jump to the wrong conclusion.

Visual illusions work in a similar way. Check out this elephant:

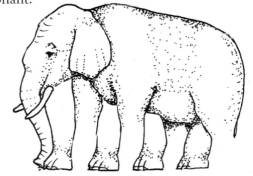

Roger Shepard's
'Impossible Elephant'

How many legs does it have? Where are its feet? Confusing, isn't it? That's because your brain already knows how elephants' bodies are put together. When it sees feet, it predicts there will be legs above them. And it knows legs should have feet below them. But in this sneaky drawing the legs and feet don't match up, so your poor brain doesn't know what to believe!

By the way, you have way more senses than the 'famous five': sight, hearing, taste, touch and smell. For example, there are senses that tell you about balance, appetite and where your body parts are (called 'proprioception'). All of these senses work in a roughly similar way. Your brain makes top-down predictions which are then cross-checked against signals coming in from the senses. Your brain only lets you experience the bottom-up data if it does **not** match the top-down prediction.

And because they work this way, most senses can be 'tricked'. For example, you can convince a friend that their nose is **as long as their arm**:

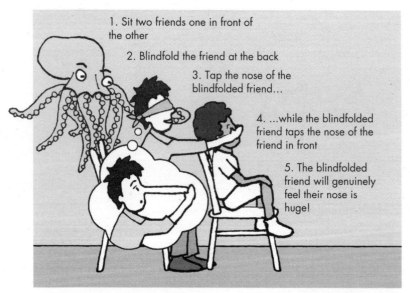

1. Sit two friends one in front of the other

2. Blindfold the friend at the back

3. Tap the nose of the blindfolded friend...

4. ...while the blindfolded friend taps the nose of the friend in front

5. The blindfolded friend will genuinely feel their nose is huge!

This 'sensory illusion' was invented by a neuroscientist called V. S. Ramachandran to show what happens when top-down and bottom-up predictions clash. Your brain usually learns from its mistakes and comes up with updated and (hopefully) better top-down predictions, but in this case the prediction was a bit off: your friend's nose had not in fact grown massively long.

Ha! You lot are so easy to fool!

Well, most neuroscientists suspect your brain works in a similar way, Octopus – cross-checking top-down predictions with information from the senses. And they think our brains work this way for three very good reasons:

1. **It saves a LOT of energy.** Your brain makes up just 2% of your body's weight, but it hoovers up almost a quarter of the energy you get from your food each day.

I can't RUN any more, I'm THINKING too much!

Since everything the brain does 'costs' so much energy, your brain does what it can to reduce the number of complex operations it has to perform. Using the past to predict your future, and only focusing on unexpected details (i.e. differences between top-down and bottom-up predictions) turns out to be a lot less effort for your brain than constantly having to analyse every last shred of evidence coming in from your senses.

2. **Your brain can use its past experiences to understand the present.** That's how you figured out what the jumbled sentence meant on page 78. It's also how you recognize faces, songs on

the radio and the smell of fresh-baked bread. Your brain is constantly blending what's happening now with memories.

3. **Awesome flexibility**. As the 'massive nose' illusion proved, our brains are usually willing to revise and improve top-down predictions that don't fit with bottom-up predictions to help us adapt to new situations.

Of course, sometimes our brains get 'lazy' and rely too much on their top-down predictions. Because we base our predictions on things we've read, seen and heard in the past, we can end up 'stereotyping'* or jumping to the wrong conclusions about people, for example. This also explains why we sometimes see faces in random places, e.g. inside clouds:

But your brain does its best with the experiences that you feed it each day. And in order to make the best, most useful predictions it possibly can, your brain spends an awful lot of time – well, a lifetime, in fact – learning all it can about the world. Which is what we turn to next.

*Stereotypes are a type of *implicit memory* (see p. 67) that our brains form and recall automatically, without us even thinking about it.

CHAPTER 5
How Brains Change Themselves
YOUR FLEXIBLE BRAIN, FROM CRADLE TO GRAVE

Are you the same person you were when you were a baby?

On the one hand, the answer is a definite 'yes'. You share so much with that tiny wriggling suckling version of 'you': a body and (biological) parents, maybe some siblings, a life history, a name and early memories, for example.

But on the other hand, you are radically different. And it's not just the obvious stuff, like being bigger, faster, hairier (possibly) and more in control of your limbs and bowels. Compared to your baby self, you can feel more complex emotions and express yourself in all kinds of fancy ways. The biggest differences between baby 'you' and today's 'you' are the result of all the changes that have been piling up inside your brain ever since.

These brain changes are caused by two main factors:

1. **Genes** – the 'instructions' that tell your brain cells how to grow, develop and wire themselves up. Different *gene* instructions are activated at different stages of life, changing the inner workings of your brain.

2. **Environment** – more than any other part of your body, your brain's development is shaped by your surroundings – biologists call this your environment. That includes everything from the places you've lived, to the people you've hung out with, to the things you've done and even the thoughts you've had. All these things have the power to alter your brain.

Getting stuck in this book is definitely gonna change my brain for ever!

But the split between 'genes' and 'environment' is never clear-cut, not least because our genes build brains that are **specifically designed** to learn everything they can from their environments. Brain changes, therefore, almost always involve genes **and** environment.

Neuroscientists call the brain's ability to shape itself according to what's happening all around it *neuroplasticity*. 'Neuro' means to do with the nervous system, and 'plasticity' means mouldable, like **plasticine**.

Because of neuroplasticity, your brain is changing during every single moment of your life. And this matters because, when your brain changes, YOU change.

> Reeally? But I know I'm still the same person I was yesterday!

Well, most 'neuroplastic changes' are tiny and hard to detect – they might involve adjusting a couple of your brain's 600 trillion synapses to form a memory (see pp. 66-69). But over weeks, months and years, loads of little changes add up.

Bigger 'neuroplastic changes' are also possible, though. For example, if an accident or illness leaves someone *visually impaired*, and they learn to read Braille and navigate with a cane, large parts of their brain that used to handle vision (see p. 20) will switch to processing touch information instead.

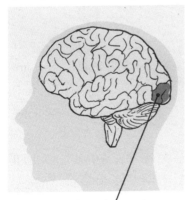

In sighted people, visual cortex mainly processes sight

In visually impaired people, visual cortex often switches to processing other senses. So, neuroplasticity means they can 'see' touch, sound and smell

Neuroplasticity doesn't happen at the same pace throughout our lives. It's more dramatic during some periods than others, depending on what our brains need to do. To see this in action, let's meet some members of a typical family . . .

My family - 'typical'? Ha!

. . . and peer inside their heads to see some of the biggest changes that are underway.

MEET THE FAMILY
Baby Bo (one month old)

Believe it or not, Bo's brain contains more neurons than YOUR brain does right now!

Bo

But Bo can't even walk or talk! How can that be true?

Yeah, she's useless! Baby octopuses go hunting as soon as they hatch!

Well, we're each born with almost all the neurons we're **ever** going to have. Lots of other animal babies, including octopuses, arrive with nervous systems wired for action, but human baby brains still have a huge amount of developing to do.

Brain size: Only about a third of the size of an adult's brain, but growing fast: loads of new glia cells are being born, nerve pathways are being wrapped in myelin sheaths (see p. 30) and neurons are busily reaching out and forming new synapses.

Brain superpower: Adaptability. Bo's brain has one particularly massive problem to solve. As we know, our brains try to keep us alive by predicting the future, based mainly on the past (see Chapter 3). But Bo hasn't got much past experience to work with, so her brain is hungrily learning everything it can. It's particularly crucial that she forms strong relationships with other human beings – especially close family members – who can help her learn and grow.

Bo watches mum carefully and even tries to copy her facial expressions

All this development and neuroplasticity taking place **after we're born** is what makes the human species so adaptable. If you sent Bo off to live with **another human family** anywhere else in the world – from frozen north to parched desert or tropical forest – she'd grow up speaking the local language and adapt to everything else just fine.

مرحباً، ان اسمي بو

Hello, I'm Bo!

Dearvva. Mu namma lea Bo.

Kim (2.5 years old)

Thanks to an awful lot of copying, loads of practice and masses of play, Kim has developed and strengthened the neural circuits that let him talk, walk, dance, romp about and cause all manner of mischief.

But Kim can throw an impressive tantrum . . . As with all toddlers, his brain's limbic system, which deals with strong emotions (see p. 21), is more fully developed than the reasoning and listening parts of his cortex, so he can't control those emotions or think about other people's needs.

Brain size: 25% smaller than an adult's, but absolutely jam-packed with synapses – two million of which are forming every single second! In total, Kim's brain contains one quadrillion (one thousand trillion) synapses, which is about **twice as many** as a typical adult.

Brain superpower: Learning. With all those brand-new synaptic connections available, Kim's brain is like a sponge, especially for language. Broca's area (see p. 43) is growing fast, as he picks up a few new words or phrases every day. And as the neural circuits in his motor cortex (see p. 20) and cerebellum (see p. 18) get stronger, he's constantly learning new skills, like 'painting'.

Ren (10 years old)

Someone pass this kid a pencil, cos her brain is fizzing with ideas and overflowing with stories!

Brain size: 90% the size of an adult's. But from now on, the total number of synapses in Ren's brain will always be falling. That's because much of the learning we do involves destroying synapses that aren't doing anything useful, while strengthening the ones that are. It's a bit like trimming back a massively tangled and overgrown plot of land to turn it into a beautiful garden, which is why this kind of neuroplasticity is called *synaptic pruning*.

Brain superpower: Some neuroscientists think this age is a sweet spot for **creativity**. Ren has learned a lot about the world – from school, friends, family, books, screens and play – but she isn't bogged down by notions of how things 'should be' done yet. Meanwhile, her glia cells are busy adding myelin to speed up nerve connections between various parts of her brain, explaining why Ren makes surprising links between sights, sounds, feelings, memories and more:

Once, in a village where it snowed every Tuesday and the trees sang to the birds, there was a mystery that got everyone talking . . .

Jay (16 years old)

Here's Jay's plan: skateboard down the roof, catch some 'mad air' off the gutter, then '360' on to the trampoline.

I can't watch!

Yep, Jay's a daredevil, but he still cares about what people think. When the video of this stunt goes up on social media, Jay will be desperate to get a ton of 'likes'.

Brain size: Fully grown, but until about 25 years ago, neuroscientists thought brains had pretty much stopped developing by Jay's age. Since then brain scans have revealed that loads of big changes are still taking place during the teenage years. Jay is attracted to risk partly because of what's going on in his *reward circuits* – which use a neurotransmitter called *dopamine* to trigger 'rewarding' feelings when we do new or pleasing things. Jay's brain is highly sensitive to these signals right now. Combine that with the fact that 'self-control' circuits in the front part of his cortex are still developing and it's no surprise that teens like Jay enjoy pushing the limits.

Brain superpowers: *Empathy.* The flip-side of caring about what others think of him, is that Jay has recently become much better at figuring out what's going on inside other people's brains. That's mainly because of newly strengthened connections between various parts of the cortex and the limbic system (see p. 21).

Mum (41 years old)

Life is never dull in this household. Someone's always tired, bored, hungry, late, cross, needing the toilet, wanting a lift, etc. On top of all that, Mum's writing a book. Yet, somehow, she never (well, almost never) loses her rag.

Brain size: Slightly smaller than Jay's (but don't tell Jay, cos it'll only go to his head!).* That's right, after the age of around 30, the human brain starts to shrink ever so slowly. It doesn't mean Mum's losing her edge, or that she can't learn new things, but it's just not quite as easy as it once was.

Brain superpower: 'Executive functions'. The self-control, planning and decision-making circuits in the front of her cortex are now in tip-top working order. That's why Mum's so good at managing 'big feelings'– her own and everyone else's!

*Male brains are, on average, also slightly larger than female brains. But, according to the science, this does NOT mean males are more intelligent than females (there's more on brain size in Chapter 9).

Grandpa Derek (82 years old)

Oops. Grandpa's just put his teabag into the toaster instead of the teapot ...

Brain size: 20% smaller than Jay's brain now, but that's totally normal for someone this age. Grandpa is, however, getting a bit vague and forgetful. His doctor says it's the early signs of *dementia*, a general term for brain diseases that kill off neurons faster than usual. Sadly, there's no cure for dementia yet, so he is likely to get worse in years to come.

Brain superpower: Wisdom. Although Grandpa's short-term memory isn't what it once was, he's still got a lifetime of long-term memories at his fingertips. So, if you're in need of some wise advice, he's the person to ask.

The Changing Brain

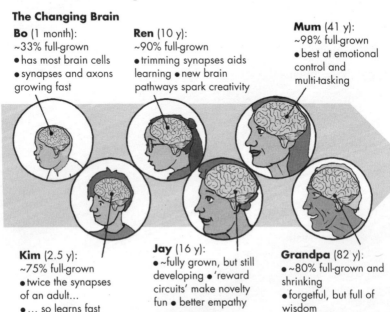

Bo (1 month):
~33% full-grown
• has most brain cells
• synapses and axons growing fast

Ren (10 y):
~90% full-grown
• trimming synapses aids learning • new brain pathways spark creativity

Mum (41 y):
~98% full-grown
• best at emotional control and multi-tasking

Kim (2.5 y):
~75% full-grown
• twice the synapses of an adult...
• ... so learns fast

Jay (16 y):
• ~fully grown, but still developing • 'reward circuits' make novelty fun • better empathy

Grandpa (82 y):
• ~80% full-grown and shrinking
• forgetful, but full of wisdom

There's a world of difference between the brains of babies like Bo and wise elders like Derek. And, thanks to neuroplasticity, the unique path each of us takes through life shapes the way our brains – and therefore our minds – turn out.

I **wish** I wasn't so **shy** . . . Maybe neuroplasticity could help me mould my brain so I can be a **totally different** person?

Not really. Neuroplasticity is powerful, but it definitely has its limits. That's because our brains need to walk a bit of a tightrope between being **flexible** enough to:

• Pick up new skills and form new memories.

• Adapt when our worlds change, e.g. when moving house, switching school, gaining or losing a family member.

• Try to heal or adapt to injuries and illnesses that damage the brain.

But also **rigid** enough to:

• Hang on to memories and skills.

• Make sure the neural circuits most crucial for keeping us alive – e.g. those that control heart rate, digestion and breathing – don't change or stop working.

• Keep our personalities and our sense of who we are fairly constant, year after year.

It's impossible to reinvent yourself completely, because neuroplasticity can't run wild. Here's why:

1. **Genes**. As we already know, genes play a critical role in brain development and neuroplasticity. However, they also limit the amount of change that can take place. Each of us inherits a unique set of genes from our parents, which builds a brain that sees, thinks and reacts to the world in its own distinctive and fairly inflexible way. Think of a building: it's easy to re-paint the walls or change the furniture, but it's much harder to change the basic structure.

2. **Capacity**. You might have heard people say that we only use 10% of our brain at any one time. That's simply not true! It's a case of '**use it or lose it**': neurons that aren't involved in active neural circuits tend to either get killed off or wired into a different circuit. In short, almost every part of your brain already has a purpose, so you can't just go on picking up new skills, habits and memories for ever, without losing some other capabilities.

3. **Age**. Generally speaking, our brains get less adaptable with age. For example, babies are occasionally born with a condition called *hydrocephalus*, which destroys large parts of their cortex. Amazingly, some people with this condition grow up to live totally 'normal' lives, without even knowing big chunks of their

brains are missing. That's because babies' brains often have an incredible ability to rewire themselves. Massive damage to the cortex later in life, however, is usually impossible to recover from.

4. **Effort**. If you were to scan someone's brain before and after they trained as a professional violinist, you'd see that the part of their motor cortex in charge of their fingers would have grown bigger. But that's only because they've put in thousands of hours of practice.

So if you lot tried hard enough you could stop being so big-headed?!

Change is always possible, Octopus, but bigger shifts usually require a lot more effort!

Neuroplasticity means our brains – and therefore our selves – are always, slowly but surely, **changing**. But because it provides a way for memories to take root and grow, it's also a big part of the reason we feel like the same person, day after day, year after year.

You mean neuroplasticity makes me, me?

Spot on. But, the question is, do you actually know what you mean by 'me'?

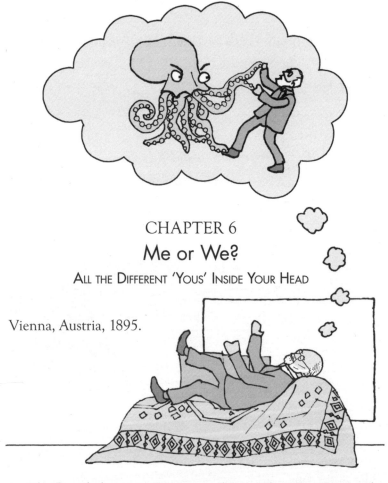

CHAPTER 6
Me or We?
ALL THE DIFFERENT 'YOUS' INSIDE YOUR HEAD

Vienna, Austria, 1895.

The bearded guy snoozing on the couch is Dr Sigmund Freud. During the late nineteenth and early twentieth centuries, he helped invent psychotherapy (see p. 130) and came up with a revolutionary new way of understanding our brains.

He doesn't exactly look like a genius . . .

Yeah, and why's he writhing around like an octopus?!

Shhh, don't interrupt Dr Freud. Dreaming is a crucial part of his work.

Freud was convinced that our dreams contain all sorts of clues that could help solve the huge mystery at the very heart of all our lives . . .

Who am I? And why am I like this?

Freud trained as a neurologist, but soon decided that the best way to comprehend human minds was by paying extremely close attention to every detail of the thoughts, feelings, ideas and actions that spring from our brains – even when we're asleep. He spent years studying his own thought patterns and scrutinizing everything his patients said during psychotherapy sessions.

When Freud weighed up all these experiences, he reached one enormous – and genuinely world-changing – conclusion. People found it deeply shocking 130 years ago and you might still find it shocking today.

Your conscious mind is NOT always in control.

Eh? Why's that so shocking??

Freud is talking about your *consciousness*: the thing that 'wakes up' in the morning and then spends the entire day 'experiencing' a non-stop stream of thoughts and sensations. Does it ever feel like there's a 'mini you' sitting inside your head, peering out of your eyeballs – a bit like Descartes's rational soul from page 39?

That's me! I'm your consciousness!

It might feel as if that conscious part of your brain is pretty much in control of your world – steering your body around and weighing up decisions about where to go and what to eat, wear and say, etc. – a bit like a pilot flying an aeroplane (**without** *autopilot*).

Pah! It's not like that at all!

OK. We'll dig into what consciousness actually 'is' in the next chapter. But, based on his observations, Freud argued strongly that if your body really is like an aeroplane, it is **mainly** controlled by the 'autopilot', not the human pilot. Using brain systems we're not even remotely aware of, your autopilot carries out millions of essential, yet fiendishly complicated, day-to-day actions so you don't have to – for example – deliberately choose which emotions to feel, decide where to place each footstep, remind your heart to beat or tell your eyes where to look.

But the part people found most unsettling was Freud's claim that the brain's *unconscious* autopilot systems frequently **ignore** the conscious 'pilot'. He said these systems often had their own ideas about where the 'plane' should go, which dangers it should avoid and how fast it should fly.

This is neuroscientist David Eagleman. Brain science has come on in leaps and bounds since Freud's time, and although some of the details of his theories turned out to be wrong, many neuroscientists, including Eagleman, think Freud's big idea was spot on: your consciousness is just one small part of the mind you call 'you'.

Remember the conversation in your head right at the start of this book (see pp. 7-8) – with all those squabbling 'voices'?

Yep. According to Eagleman's theory, each of those 'voices' represents a different aspect of 'you'. We can think of all these different 'yous' as a **team** inside your head, because they all share the same basic goal: to keep your body alive and well. But they're also **rivals** because they've got clear ideas about how to achieve that goal – and they're all convinced that their way is the best. We'll head back to that conversation on page 102.

Meet Your Team

Those 'debating' voices were some of the loudest members of your 'team'. Here each voice is shown as a character that represents a different neural circuit (see p. 31) or part of the brain normally involved in one particular way of thinking:

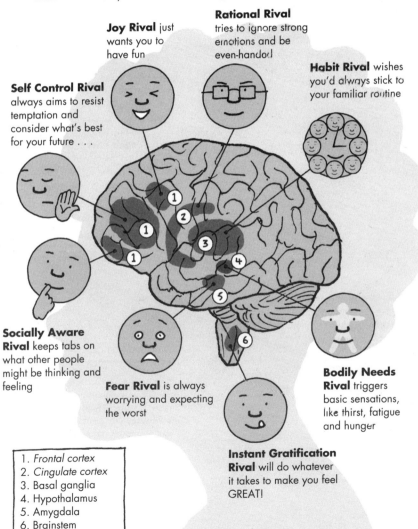

Here's the conversation again, but this time we'll see where the different comments are coming from:

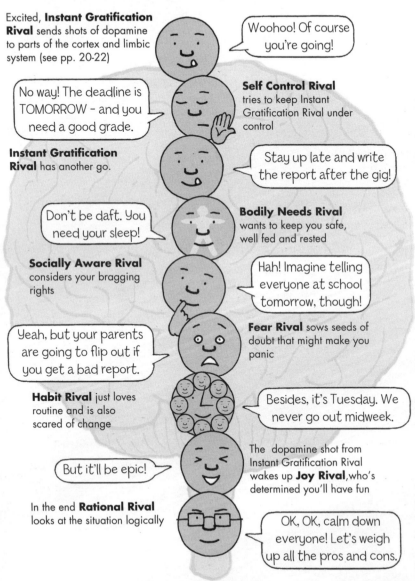

Too true. There are loads of other brain systems jostling to influence you. For instance, a whole array of 'emotional rivals' – both pleasant (joy, love, gratitude) and less pleasant (jealousy, disgust, embarrassment) – inhabit various parts of your limbic system and cortex.

None of our inner rivals are totally separate, of course. They all belong to the same nervous system and they're all communicating all the time.

> Which is why you can get the weird sensation of arguing with **yourself**!

Absolutely. And although we've given them 'voices', most rivals don't literally use words. Instead, the rival brain circuits make their points by releasing squirts of neurotransmitters (see pp. 28-29) and hormones and changing which neurons get activated. These changes then stir up strong feelings, emotions or memories – often without your conscious mind having **any clue** where they're coming from.

The 'rivals' react differently depending on what else is going on in your life – like how hungry, tired or stressed out you are. If you had a really bad night's sleep, for example, your 'Self Control Rival' might lose the battle and you'll impulsively decide go to the concert, whatever the consequences.

> So which of these rivals is the real 'me'?

All of them and none of them! In other words, there's no single 'ideal' version of 'you'; you really can be a different person at different times and in different situations – at home, at school, on a first date, etc.

> But why does it have to be so complicated?

Well, having different rivals (or brain systems) with different strengths and weaknesses all trying to crack problems in their own way means no single thought process is in control.

> Hmm.

Imagine 'Rational Rival' was solely in charge. Without any emotional urges to steer your choices, even fairly simple decisions, like what clothes to wear, would leave you dithering while your brain tried to weigh up every scrap of evidence and all possible consequences.

Now picture what might happen if 'Instant Gratification Rival' was the boss. You'd be bouncing around like an over-excited puppy, impulsively jumping from one hasty choice to the next, as your brain constantly chased after whichever option looked most fun or most satisfying in the moment.

Having a whole team of rivals inside your brain is far from perfect, however, partly because the various rivals evolved gradually, over millions of years. In fact, the basic design of our human brains has barely changed since around 50,000 BCE, during the Stone Age.

Not exactly. But an awful lot has changed in the last fifty thousand years, and our 'Stone Age' brains aren't always perfectly suited to today's buzzing, interconnected world. For example:

In 50,000 BCE, sugar would have been a very occasional

treat, so it made sense to give in to the hunger triggered by Instant Gratification Rival.

Today we're often surrounded by sweet treats, so we have to work to keep Instant Gratification Rival under control . . .

And in the Stone Age, physical danger was all around, so you needed 'Fear Rival' to make you drop everything and run for your life. These days, the same rival can sometimes get confused and overreact:

For most of us, however, life really is safer, easier and much more pleasant now we're no longer living in the Stone Age. We have amazing healthcare, awesome entertainment, mind-boggling technologies and democracies where everyone gets some say in how their country should be run. The reason we've been able to create all this is that people have found ways of getting the 'teams of rivals' inside their skulls to work together for the benefit of themselves and everyone else.

We can all, at least on our good days, rise above the bickering, small-minded and selfish parts of ourselves and decide to behave with enormous kindness, extraordinary cooperation and exceptional ingenuity.

If Freud was still around, what would he make of this modern view of the brain?

Your fancy new brain scanners are mainly confirming stuff I figured out over a 100 years ago!

And Freud, of course, was the first person to point out that our brains do most of their work – controlling our bodies, making predictions and steering our decisions – silently and automatically, while our conscious minds hardly need to get involved. That doesn't mean your consciousness doesn't matter, however. Far from it; if it wasn't for your consciousness you quite literally wouldn't think or feel anything. You wouldn't really 'be' anyone.

Whoa. So what, exactly, is this consciousness thing?

Big question.

Big enough to get its own chapter.

107

CHAPTER 7
How It Feels to Be You
The Consciousness Conudrum

What's it like to be an octopus?

Well, we've got eight arms – each with their own brain – so it's pretty awesome.

Hmm . . . with our **empathy** superpower, we can **try** to **imagine** how that might feel.

And we can see and taste through our incredibly sensitive skin!*

Woah! Now you're just blowing my mind!

Sweet!

Ooh, sour!

Mmm, orangey.

*See page 64.

But we'll never **know** what it really feels like to **be** an octopus, or even another human. This is the mystery of consciousness. You know exactly what it's like to be **you**, but you'll never be able to truly understand how it feels to be any other living being. Your consciousness is the most private thing in the world: a show put on by your brain just for you.

The inner experience of every other human being might be totally different from yours.

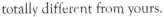

Even if two different people agree that the colour blue definitely exists, it's impossible to know whether they both 'see' blue in exactly the same way.

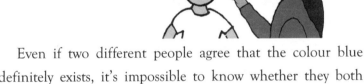

Nope. Some people are born with a dangerous condition called '*congenital* insensitivity to pain'. To them, a hand on a hot pan feels no different from a hand in ice-cold water. Their brains just don't produce the sensation of pain.

Not necessarily. When people have a condition called *synaesthesia*, their senses get blended. They might taste colours, for example, as well as see them; or see music as well as hear it . . . and sugar might taste sweet, but also sound like violins.

That may be an extreme case, but since all our perceptions are 'cooked up' inside our brains (see Chapter 4) – and our brains are all different – it makes sense that we each experience the world a bit differently. But the million-dollar question is, who or what is having these experiences? In other words, what is consciousness and where does it come from?

Could it actually be anything like René Descartes's rational soul (see p. 39)? Is there really a 'mini you' inside your head?

Well, sorry to have to say this 'Mini You', but you don't actually exist.

Well, saying there's a little conscious 'self' inside your head just makes consciousness even more complicated. If there were a 'mini you' (or a rational soul) in your brain, we'd need to find out how **it** was conscious. Would the 'mini you' have a 'mini-mini you' inside **its** head?

And, the 'mini-mini you' have a 'mini-mini-mini you' . . .

Instead, most neuroscientists generally think of our brains as super-complicated, electric-powered, information-processing machines made from flesh and blood, that have somehow 'woken up' and become aware of their own existence. One big reason scientists believe this is that when they change the pattern of electrical activity going into our 'brain machines', the behaviours and **conscious experiences** the machines produce change too.

When 16-year-old Anna K was undergoing treatment for *epilepsy*, brain surgeons inserted a series of small electrodes into various parts of her brain to help pinpoint the cause of her seizures.

After the operation, Anna's doctors talked to her and showed her pictures, while activating the electrodes one at a time. Every time a particular electrode in the top part of her cerebral cortex was turned on, Anna started giggling uncontrollably. Flicking that electrical 'switch' inside Anna's 'brain machine' suddenly altered her conscious experience.

But how can our physical brain circuits create our own private feelings of consciousness?

The honest answer is that nobody knows.

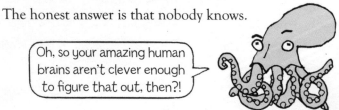

Oh, so your amazing human brains aren't clever enough to figure that out, then?!

Unfortunately not. How the brain generates consciousness is one of the deepest, most challenging mysteries in all of science.

That's why I call it the Hard Problem!

Yep, that's how philosopher and brain scientist David Chalmers summed things up, back in 1994.

EMERGING ANSWERS TO A HARD PROBLEM

Many neuroscientists think the answer to the Hard Problem is that consciousness is an *emergent property* of the brain. In other words, individual brain circuits don't have a consciousness, but when many of them work together they somehow create conscious awareness.

To see how this might work, think of ants. One ant on its own definitely doesn't have the strength or knowledge needed to build an anthill. But put thousands of ants together in a colony and they each start gathering balls of earth and piling them up. Eventually, incredible structures – like arches, tunnels and ventilation systems – take shape!

No single ant plans it, but together they create something complex and brilliant. The whole colony is far smarter than any one ant.

Lots of scientists think consciousness is a bit like that ant colony, only built from **billions** of neurons and glia, rather than **thousands** of creeping ants. And, just as you can't fully understand an ant colony by understanding a lone ant, neuroscientists will never find consciousness in any single brain region or circuit. Instead, their challenge is to find out how lots of brain processes work together to create conscious experiences: the emergent properties of whole brains. Researchers are working their socks off to test this theory, but they're not there yet.

And they might never get there!

Really!? A human who doesn't think he knows it all!

Well, we don't. It might take something even more complicated than a human brain to fully understand a human brain.

So we can't say for certain **how** consciousness happens, but some neuroscientists have firmer theories about **why** it exists.

THE HEADTEACHER THEORY

One popular idea is that our human kind of consciousness is a bit like the headteacher at a busy school. Often it's other teachers, caterers, caretakers, office staff, etc. who keep the school running. You might not see much of the head at all . . . until something goes wrong.

Then the headteacher leaps straight into action, making sense of what happened, deciding what to do and explaining everything to the school inspectors, parents, etc.

In the same way, you have a set of brain systems that keep a general eye on what's going on in and around your brain and body. They're not crucial for your body to function, but they leap straight into action whenever something surprising happens, e.g. if there's a serious mismatch between your brain's 'top-down' predictions and the data coming in from your senses (see Chapter 4). Or when you've got a tough decision to make.

Like whether to go to an amazing gig or do your urgent homework project?

Exactly. We're back to that dilemma from page 7 and your team of rivals, from pages 101-2.

So to make that decision my inner 'headteacher' has to get involved, right?

Yep, that's the idea. According to the headteacher theory, your consciousness 'listens' to the various inner rivals battling it out inside your head, chooses the rival **it thinks** should win, then does it's best to influence the result.

Your consciousness doesn't always make the right decision,

though. Not least because there's loads happening in the 'school' (your brain) that the 'headteacher' (your consciousness) can't understand or control – e.g. all those unconscious brain processes that are silently whirring away, day and night, to keep your heart beating, senses working and strong emotions, ideas and memories 'popping up'.

And sometimes it'd be better if your consciousness **didn't even try** to get involved.

Once you've mastered a skill, like playing tennis, for instance, it's usually better to let your unconscious, implicit memory systems (see p. 67) take charge of the game.

Tennis star John McEnroe, who won over seventy major tournaments during the 1970s and 80s, clearly knew this. When his opponents were serving the ball really well, for example, he would congratulate them.

WOW – EPIC SHOT!

They'd then become conscious of their service, and often start making mistakes!

Another intriguing thing about your consciousness 'headteacher' is that it likes to feel as if it's in control even when it definitely isn't. And it can weave a great story, even when it doesn't actually know the truth. For example, when a doctor asked Anna K why she'd suddenly started giggling during her treatment (see pp. 111-12), she had no idea her brain was being stimulated by an electrode. But her conscious mind quickly came up with a believable explanation:

So what would happen if we didn't have a consciousness to steer our actions?

A herring gull, for example, doesn't have an equivalent way of choosing between rival neural circuits. Seeing a red object triggers aggressive behaviour; seeing an egg triggers caring behaviour. Place a **red egg** in a herring gull's nest and the poor bird will lose its mind trying to attack the egg and incubate it at the same time!

Of course, we humans often dither and struggle to make our minds up too. But, compared to herring gulls at least, our kind of consciousness seems to give us more options for thinking things through.

You're probably right, Octopus. Consciousness wasn't suddenly turned on like a lightbulb as humans were evolving.

When Did Life 'Wake Up'?

Think about your own consciousness. Does it feel different when you're wide awake from when you're sleepy? Or when you're dreaming rather than 'out cold' in a really deep sleep. These are all different states of consciousness. Other animals probably experience their own distinctive forms of consciousness too.

Remember the six stages of brain development (see pp. 59-62)? Some scientists think simple forms of consciousness may have emerged quite early in that process, perhaps way back at Stage 2 or 3. They argue that consciousness evolved to help brains make better predictions about threats and opportunities, so they could then choose the best way to react. Here's how that **might** have happened.

In order to work as a 'control centre', early nervous systems had to monitor what was going on in an animal's body. To do this, brains started storing an 'inner model' of how the body should ideally be, e.g. normal body temperature, heart rate, breathing rate, expected sensations from muscles, energy levels, etc.

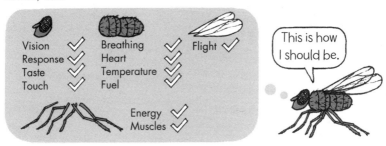

The animal could then compare its inner model to the current state of its actual body.

If there was a mismatch – e.g. heart beating too fast or a lack of energy detected – the brain would know that it needed to act fast to make its real body match its ideal inner model.

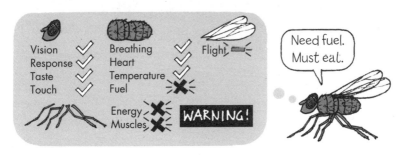

The first mismatches might have been the first forms of conscious experience – sensations like hunger, thirst and fear. Although our human brains have done a lot of evolving since, these basic sensations are still at the core of many of our conscious experiences today.

Over time, the brains of some species expanded their inner models to include maps of the outside world too, helping to predict where to find food, or what a predator looks like, for instance.

As animal minds got more complicated, some species' inner models began to include maps of the brain itself. This could be how brains first developed forms of consciousness that could listen to and influence their own internal team of rivals.

Somewhere along the way, some animal minds developed the amazing ability to empathize – i.e. step into another individual's shoes and guess what kind of private experiences they were having.

It doesn't matter: empathy is the power to **imagine** what any other being **might** be feeling, including what it's like to be an octopus. Scientists suspect that a surprisingly wide range of animals, including mice, dogs and pigs, might be capable of some degree of empathy.

In 2023, researchers did an experiment to see what would happen when a pig was separated from its group, and shut into a small pen on its own. Whenever this happened, the pig's pen-mates responded to the unhappy pig's calls and tried hard to help it escape. It really seemed they were sharing the trapped pig's distress.

You're absolutely right, Octopus. Now that humans know how much other animals can feel, we really should be using the latest discoveries from neuroscience to make sure we treat pets, farm animals and wild animals, like you, with the care and respect you all deserve.

So how come it was humans who figured it all out?

It's all down to the way our particular form of consciousness developed. As our ancestors evolved to become human beings, our conscious powers just kept expanding. We began to use language, to dream up stories, make art, invent tools and swap knowledge.

Which is why I said, '*your brain is wider than the sky.*'*

Exactly! Our inner models grew massively, to incorporate maps of our entire worlds and everything we know.

Eventually, people worked out how to do neuroscience – in other words, we started to use our conscious minds to decipher the inner workings of our own brains. Which is why we now know for certain that, just like snowflakes, every single human brain is unique and beautiful.

So you're also as fragile as a snowflake, right?

No! Thank goodness. But it is really important we take good care of our brains – and the brains of the people around us. Life can quickly get miserable if we don't.

So how do we look after them?

Let's find out.

*See poem on page 10.

CHAPTER 8

Mind Your Head

MAKING THE MOST OF YOUR ONE AND ONLY BRAIN

Your brain is unbelievably complicated. It assembles itself from a stupendously massive number of squishy, fragile parts – none of which can be seen with the naked eye – and it never stops changing. Because of neuroplasticity (see Chapter 5), your brain is constantly reacting to everything that's going on in and around it.

There are over **eight billion** human brains alive right now, and each of them is a genuine one-off. And since our **brains** are all different, **we** are all different.

Our innermost thoughts and feelings happen inside the privacy of our own skulls, so it's easy to forget that the same situations can affect the people around us in ways that are sometimes similar and sometimes **completely different**. As this is a book, however, we can peer into people's minds and see which thoughts, feelings or 'brain rivals' (see pp. 101-102) are currently in charge in this typical school classroom.

It's a history lesson! We're all bored, right?

Hmm, you might be surprised . . .

- **Alice** is terrified because of the tiny spider crawling across the ceiling, but she's trying not to show it.
- **Jess** can't stop thinking about Carter. He's so amazing.
- **Jo** is really, really missing her pet dog.
- **Carter** is feeling anything but amazing. He's spotted his reflection in the window – and hates what he sees.
- **Soly** loves history, but he's really frustrated because the words on the whiteboard seem to keep swirling into an unreadable mess.
- **Benedict** has raced through all the teacher's tasks and is now wondering whether they'd rather be called Aza or Ace.
- **Maya's** been told off – again – for day-dreaming. But she can't help it; she's a fidgety ball of energy today, constantly distracted by ideas that keep zipping through her mind.
- **Kira's** doing her best to keep up, but she didn't eat breakfast, so her mind is dominated by hunger.
- **Obie** has no problem focusing. But, at the back of his mind, he's dreading the end of the lesson: he finds noisy crowds and bustling corridors really tricky.

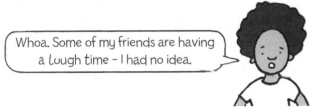

Whoa. Some of my friends are having a tough time – I had no idea.

Being a human isn't easy. We all have our own strengths and challenges, loves and hates, hopes and fears. We all have periods when we're feeling on top of the world and periods when we're feeling truly rotten. Most of the time we're somewhere in between.

And just as we can't look after our physical health by 'deciding' never to catch a cold virus or stub our toes, we can't just 'decide' to have perfect mental health all the time.

And then there are *neurodevelopmental* conditions.

Neuro-whats?

Neurodevelopmental conditions. They can partly be caused by the genes people inherit. They alter the way brains develop, think and experience the world and can give people particular skills and abilities. They can also cause disabilities or turn aspects of everyday life into huge challenges. Some of the more common neurodevelopmental conditions include:

• **Autism-spectrum disorder (ASD)**. ASD affects the way people learn, communicate and interact with others. Symptoms vary from being barely noticeable to severely disabling. People with ASD can have specific strengths (e.g. an amazing memory or grasp of logic), as well as specific needs or challenges (e.g. difficulty expressing emotions or meeting strangers). It's possible Obie (see p. 125) has ASD.

• **Attention deficit hyperactivity disorder (ADHD)**. People with ADHD may find it hard to sit still and pay attention, follow instructions or control their emotions. Perhaps Maya has this condition.

• **Specific learning disorders.** *Dyslexia* makes reading and writing more challenging, and *dyscalculia* affects maths and number processing. Soly could have dyslexia.

How does anyone know they've got one of these conditions?

The dividing line can be very blurry. For example, Autism-spectrum disorder occurs along a 'spectrum'. People can be mildly autistic, severely autistic or somewhere in between. Experts suspect it's similar for many neurodevelopmental conditions. Being officially diagnosed as having one often simply means a person experiences the world a bit differently from most other people. Some prefer to call themselves neurodivergent, to celebrate the whole amazing range of brains that share our world. And while people with these conditions may need treatment or extra support to help with particular symptoms, not everyone **wants** treatment, and these generally aren't conditions that can be – or need to be – 'cured'.

Climate activist Greta Thunberg has ASD. She believes her condition has helped her to see the problems caused by the climate crisis extremely clearly – and take action.

Being different is a superpower.

SKOLSTREJK FÖR KLIMATET

Though even Greta would agree that her ASD doesn't always feel like a 'gift'. And for some people, having a different brain may never seem amazing. But the world would be a pretty dull place if all brains worked in the same way.

WHEN MENTAL HEALTH ISN'T HEALTHY

While some people inherit genes that make them feel life's highs and lows more keenly, mental health conditions usually have complicated causes. For example, they can be triggered by traumatic events (e.g. being in a car crash or losing a close family member); by extended periods of stress (e.g. being bullied, lonely or overworked); or by underlying neurodevelopmental conditions (see above). Mental health conditions include:

• **Depression** – a deep, long-lasting sadness that can make everything seem dark, pointless and overwhelming. It can affect energy, appetite, concentration and sleep.

• **Anxiety disorders** – Alice's fear of spiders (see p. 125) is a *phobia* – a specific kind of anxiety disorder. Some anxiety disorders can become much more serious, however, sending our minds into overdrive, spotting dangers where none exist and obsessively thinking the worst. They can activate the body's stress response – leading to a racing heart, sweaty palms and shutting down parts of the cerebral cortex, making it genuinely impossible to stop worrying.

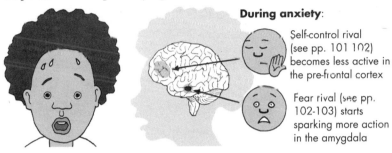

During anxiety:

Self-control rival (see pp. 101-102) becomes less active in the pre-frontal cortex

Fear rival (see pp. 102-103) starts sparking more action in the amygdala

• **Addiction** – when someone's reward circuits (see p. 90) are 'hijacked' by drugs, alcohol or particular behaviours (e.g. gambling), they experience such intense cravings for 'more' that it can become difficult to carry on living a normal life.

Mental health conditions certainly aren't anything to feel ashamed of – they're **completely normal**. According to a massive 2023 study, **half** of the world's population is likely to suffer from a mental health condition at some point in their lives. So, it's a good job help is available, thanks to the ever-growing understanding of the way our minds work.

TREATING THE MIND

Sigmund Freud developed 'talking therapies' about 140 years ago (see pp. 96-99). Today, this general approach is still helping millions of people worldwide: we call it psychotherapy. In psychotherapy, patient and psychotherapist build a special kind of relationship that can help the patient better understand and handle the thoughts, feelings and behaviours that are damaging their mental health. When it works, psychotherapy can literally change the patient's mind – thanks to the brain's neuroplasticity.

Medication – e.g. 'anti-depressants' and 'anti-anxiety drugs' – can also help with some mental health conditions. Most of them work by altering the flow of certain neurotransmitters in the brain which, in turn, alters the activity of neural circuits, e.g. the ones that control emotions or the stress response. The drugs can cause side effects, however, and, partly because all brains work in slightly different ways, they don't help everyone.*

Then there's **brain stimulation**. Ever since Aldini's early experiments (see pp. 41-42), scientists have been using electricity to try to 'jolt' people's brains into healthier states of mind. Today, using their deep knowledge of the brain, researchers and doctors can do this in extremely precise ways – and achieve some exciting results. For example, a technique called '*Transcranial* magnetic stimulation' (TMS) seems to help some people with depression or anxiety disorders even after psychotherapy and medication have failed. Here's how it works:

*Recreational – and often illegal – drugs (e.g., marijuana, alcohol, cocaine, ecstasy) also alter thoughts and feelings by adjusting neurotransmitter levels. They affect people differently and can be addictive and dangerous.

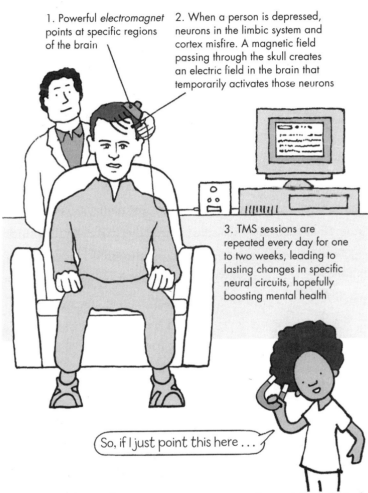

No, **don't** try the magnet treatment at home! It definitely won't work! There are plenty of other ways you can look after your brain, though, all backed up by recent research. These top neuroscientists have some great advice to share.

Do Try These Things at Home

1. Sleep

Yes! In 2012, a team, led by Danish neuroscientist Professor Maiken Nedergaard, discovered a totally new brain network called the *glymphatic system*. It's a massive web of tiny tubes made from glia cells that spreads through the entire brain. It gets busy when you're asleep, flushing out all the rubbish that builds up in your brain after a hard day's thinking.

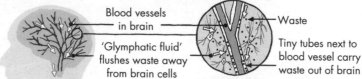

On top of that, all sorts of essential processes fire up inside your head after you drift off. For example:

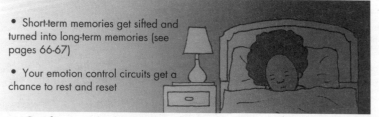

- Short-term memories get sifted and turned into long-term memories (see pages 66-67)

- Your emotion control circuits get a chance to rest and reset

Studies prove that without enough sleep, people can be less generous and more likely to suffer anxiety and depression.

Dreams can puzzle or spook us, but they're worth paying attention to. Scientists aren't totally sure yet why we dream. But most reckon dreams must matter (e.g. for processing memories and emotions). Why else would your brain work so hard to put on spectacular nighttime 'shows' just for you?

Getting enough sleep isn't always easy, though. Adolescent brains, for instance, are programmed to go to sleep later and wake up later than adults' and children's brains. Some researchers think this difference may have developed to help teenagers socialize, away from their parents.

Yay! So it's scientifically proven that I need to stay out late tonight!?

We-ell it's also a fact that you need a good night's sleep and your school day starts at nine o'clock sharp!

2. **Move!**

Your brain just doesn't work properly if you're sitting on your butt all day!

This is Dr Wendy Suzuki, one of the growing number of neuroscientists who are showing us why being more active is so good for our brains. Getting your heart beating and lungs panting can, for example, release a potent chemical called 'brain-derived neurotrophic factor' (let's just say BDNF) into the brain. BDNF acts like a 'fertilizer' for brain cells. It keeps neurons healthy, stimulates neuroplasticity and may even trigger the birth of brand-new neurons in the hippocampus (vital for learning and memory, see p. 22).

Exercise can also get neurotransmitters in the brain's reward centres fizzing, giving your mood a big boost. And the good news is that evidence shows we don't all need to be lifting massive weights or running marathons to reap the brain benefits of exercise.

If all that moving around makes you hungry, be sure to. . .

3. Eat Proper Food

Guzzle your greens and ditch the fizzy drinks!

As Professor Michael Gershon points out, a healthy, balanced diet is an important part of keeping your brain in tip-top condition. And, weirdly, he's proved it's connected with the trillions of *microbes* that live inside your intestines – your gut *microbiome*.

1. Eat a bad diet (lots of junk food, no veg) and you'll encourage unhelpful bugs to grow in your intestines

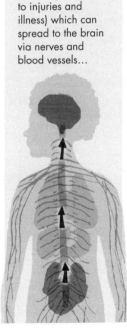

2. These bad bugs trigger *inflammation* (a normal response to injuries and illness) which can spread to the brain via nerves and blood vessels...

3. And cause depression, anxiety and even cravings for more junk food

4. Treat Social Media with Respect

Brain scientist Dr Amy Orben is a leading expert on how using social media affects our mental health.

And the effects aren't **all** bad. I've listed pros as well as cons:

Pros:
• Lots of the content is fun **and** mind-expanding.
• It can help people form friendships and find support – wherever they are in the world.
• It can provide spaces to experiment and show off your creativity.

Cons:
• It's **designed** to be **addictive**. If you get hooked, you might skip other vital activities, like sleeping.
• You might be exposed to bullying, rumour-spreading or harassment – which is seriously bad for mental health.
• You might see disturbing content (e.g. inappropriate sex or violence) that you can never 'unsee'.
• Heavy social media use could be linked to increased risk of anxiety and depression.

So, **how** you use social media really matters. Be honest with yourself. Does any time you spend on social media make you feel better or worse? Is it a pleasure or more of a pressure? And how often does it make you . . .

5. Laugh

In 2017, neuroscientists in Finland proved laughter can release powerful neurotransmitters called 'endorphins' into the brain, busting stress, dulling pain and bringing intense happiness. Professor Sophie Scott's research also shows . . .

Laughter is the glue that holds us together!

Yep, laughter makes it easier to form new friendships and make stronger bonds with old friends. We often laugh during conversations without even realizing it.

Loneliness can cause stress, anxiety and depression. But laughter is an antidote, reminding us how badly our brains **need** other people.

6. Experience Awe

Do something that makes you feel really small and insignificant!

Eh? Won't that make things a whole lot worse?

Not if it triggers an emotion called 'awe'. Professor Dacher Keltner's research shows spending time in nature – from stargazing to watching ants – can be the easiest way to experience awe. Brain scans show awe can change our thought processes in a big way. To see how, let's think back to the argument in your brain from page 7. Before Ash texts you about the gig, you're struggling with your project – your mind just keeps wandering off. That's because the 'default mode network' (DMN) in your brain is active:

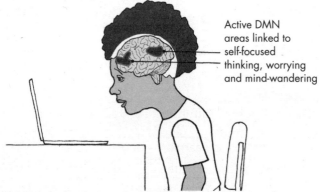

Active DMN areas linked to self-focused thinking, worrying and mind-wandering

Then Ash's message comes through, offering you a ticket to the concert. It triggers a massive debate between all your brain rivals and, in the end, you can't resist!

You're still buzzing when you get home. The awe you experienced at the gig means . . .

I can totally do this!

DMN is less active, making you curious, creative, less self-centred and more generous

You're so fired up, you finish your project in record time. Your amazing brain has come up trumps again!

Aaarghhh! You just can't help yourself, can you?

What do you mean, Octopus?

Droning on and on about human brains. You've hardly mentioned the rest of the animal kingdom – you really do think your brains are the best, don't you . . . ?

Well, it is true that scientists are discovering that all sorts of creatures have astounding mental powers. The question is, can their brains shine as bright as human brains? Looks like we need to settle this issue, once and for all!

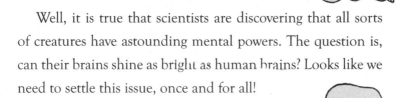

OK, big-head. Bring it on.

CHAPTER 9
All Kinds of Minds
WHALES CHAT, CROWS CHEAT, BEES PLAY AND PLANTS DECIDE

How many books have you written, Octopus?

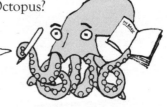

None. But so what? How many fish have you ambushed this morning, human?

Er, none. Breakfast came from the fridge. How many fridges have octopuses invented?

Who needs a fridge? I grab my food while it's still squirming.

Stop arguing! We already know that octopuses and humans are intelligent – just in different ways.

Absolutely. Intelligence is actually a tricky thing for scientists to define and measure – especially when they're comparing different species. Intelligence means using information to solve problems (see p. 56) – and every species has its own problems, depending on where and how it lives. So there are lots of different forms of intelligence, but the human sort does seem quite special. What is it about our brains that makes us able to write books, invent fridges . . .

> . . . and behave as if you rule the entire planet!?

Well, perhaps we do get a little, er, self-important sometimes, but the fact is, no other creature can use its intelligence to invent, build and alter its surroundings the way we do. We construct gigantic cities for millions of people, and keep them stocked with food, water and electricity. Our planes, trains, ships, cars, submarines and space rockets take us practically anywhere . . .

> . . . leaving a trail of destruction and stinking pollution behind you!

> Sigh. Octopus does have a point.

OK, it's true, humans create problems as well as solving them. Still, there's clearly something about our brains that makes us stand out from the rest of the animal crowd, but what is it?

BIGGER ISN'T ALWAYS BRIGHTER

It's not just about size. The biggest brains on the planet weigh a whopping eight kilograms (six times more than your brain). They belong to sperm whales.

Sperm whale brain: 8kg
Human brain: 1.3kg

Whoa! Sperm whales must be super smart!

Depends on what you mean by 'smart'. One thing sperm whales can do that most humans definitely can't is *echolocate*. They basically use sound waves to 'see' what's going on around them in the deep, dark ocean. They do it by producing streams of clicking noises that bounce off any solid objects in the water, including the whales' favourite prey – squid.*

*Some visually impaired people do in fact learn to echolocate – usually by 'clicking' their tongues and then training the visual parts of their cortex to 'see' the tiny echoes. Another example of the power of neuroplasticity.

After listening to thousands of recordings of sperm whales' clicks, biologists think they also use echolocation to have 'conversations', e.g. about where to find food or how to avoid ships. Sperm whale mums even appear to 'agree' to babysit each other's calves while they take turns diving for squid.

So what? Octopuses can communicate. Loads of animals with smaller brains than whales' can too.

True – humans included. Brain size doesn't tell us everything. That's why some scientists estimate animal intelligence by comparing brain size to body size. They call this the 'brain-to-body ratio'. The idea is that animals with bigger brains, compared with their body size, should be more intelligent.

A sperm whale has a 40 tonne body, so it's eight-kilogram brain makes up just 0.02% of its bodyweight.

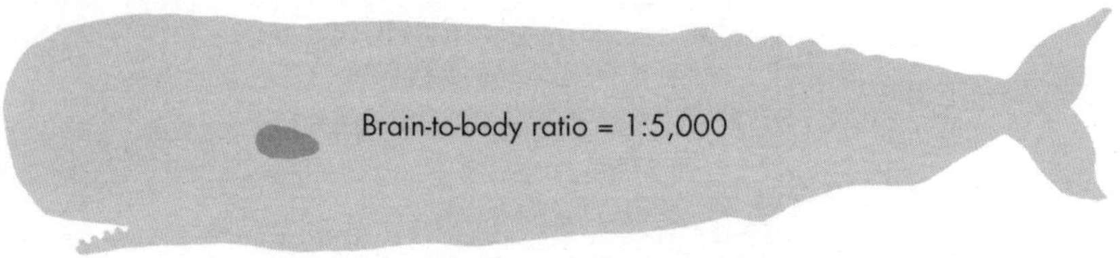

Brain-to-body ratio = 1:5,000

An octopus brain, on the other hand, is about 0.5% of its bodyweight:

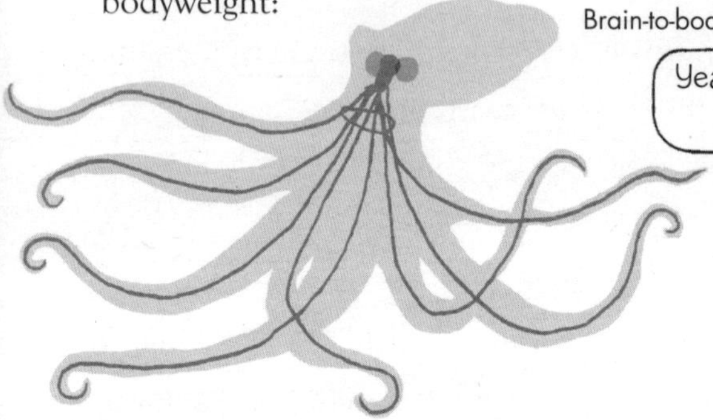

Brain-to-body ratio = 1:200

Yeah! We can outsmart a whale any day.

Don't get smug, Octopus. A human brain is 2.5% of body weight – that's a ratio five times higher than yours:

Brain-to-body ratio = 1:40

But then again, cute little tree shrews have a brain-to-body ratio that's four times higher than ours (10% of their bodyweight is brain)! Not noted for their intellectual brilliance, tree shrews are, however, among the world's smaller mammals. In order to do everything these tree-dwelling, insect-hunting mammals need to do, a tree shrew has to pack millions of normal-sized brain circuits into its relatively tiny skull. Hence, tree shrew brains take up a bigger chunk of their total body weight:

Brain-to-body ratio – 1:10

So brain-to-body ratio clearly isn't the best way to measure intelligence. Could the answer lie in the way neurons are arranged inside a brain?

CLEVER WIRING

For most birds, evolving a massive brain isn't an option – in order to fly, they need to keep their body weight low. Yet birds such as parrots, ravens and crows have developed seriously intelligent brains.

Biologists have proved that crows even have the brain power to imagine what other birds are thinking. For example, if a crow wants to hide some food for later, but thinks another crow is watching, it might change plans and **pretend** to bury the food, before flying off and hiding it somewhere else. Until very recently, scientists thought only humans could be this deceitful!

Some parts of crows' 'feather-light' brains are twice as tightly packed with neurons as they are in a human brain. That means the links between neurons can be shorter and therefore lighter.

But even jam-packing a brain with neurons doesn't explain higher levels of intelligence. The brains of honey bees contain around a million neurons – **1,500 times fewer** than a crow's 1.5 billion neurons. Nevertheless, minuscule bee brains are capable of all sorts of intelligent feats.

Honey bees plan ahead and can adjust their plans when circumstances change. For example, when a worker bee finds a great new food source – like a bed of blooming flowers – it uses a sophisticated kind of 'sign language' called a 'waggle dance' to convince all the other worker bees to help harvest it.

In 2017, a team of researchers in London, led by neuroscientist Lars Chittka, even managed to teach bumblebees to play football.

Football?

Yup, sort of. They trained a few bees to become 'demonstrator bees', by rewarding them with sugar whenever they pushed a small ball into a hole to 'score a goal'.

Other bees figured out how to play the game without any training at all, just by watching.

Amazingly, these 'watcher bees' did something far more impressive than mindlessly copying the demonstrator bees:

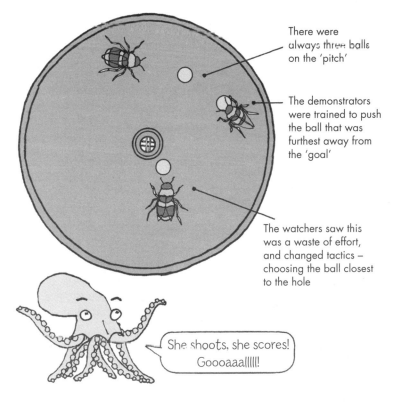

There were always three balls on the 'pitch'

The demonstrators were trained to push the ball that was furthest away from the 'goal'

The watchers saw this was a waste of effort, and changed tactics – choosing the ball closest to the hole

She shoots, she scores! Goooaaallllll!

Although he hasn't yet found a way to prove it conclusively,

Chittka suspects bees could be conscious and capable of a surprising range of sensations and states of mind, including pain and hopefulness. The challenge is on to work out how the relatively small number of neurons inside bees' poppyseed-sized brains can do so much.

Perhaps even more surprising, however, is the growing realization that living things don't need even a single neuron to solve challenging problems in intelligent ways. That's definitely true for plants.

Can Plants Ponder?

Did you know that the smell released by fresh-cut grass is actually a cry for help? It attracts predators, such as birds and wasps, that swoop in and attack any herbivores that are chewing on it.

That's a clever **reaction**, but plants can also **predict** and **anticipate**. Bees prefer collecting pollen and nectar from sun-warmed flowers, so, rather than simply reacting after light shines on them, some plants swivel their flowers to face east

before the sun has even risen. They also seem to 'remember' what time of day pollinating insects have visited in the past and use those 'memories' to release extra pollen at the exact moment it's most likely to be collected and spread around by the pollinators.

> But plants don't have brains!?

No they don't, but that doesn't stop them:

- using tiny sensors in their leaves, stems and roots to detect light, 'taste' chemicals and 'hear' sounds

- changing the chemicals in their cells, or the shape of their bodies to hang on to 'memories'

- transmitting information through their bodies with the help of hormones and electrical signals (a bit like action potentials, see p. 27).

Plant 'intelligence' may be brain-less and rather slow, but it works: how else could they have painted practically every inch of available land green?

> Maybe human brains aren't so special after all!

> You're finally getting it!

OK, so now we're back to the question we started this chapter with . . . and we still haven't found that single magic feature that marks our brain structure out as unique. We know our brains are big, that they're big compared to our body size, and that they're jam-packed with neurons, but we also know other creatures share all these features. And it's a fact that the brain of our closest relative, the chimpanzee, has all the same basic regions as a human brain, but the human brain – especially the cerebral cortex – is bigger overall.

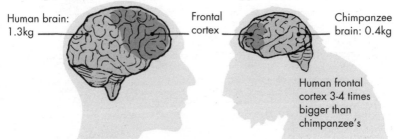

Human brain: 1.3kg

Frontal cortex

Chimpanzee brain: 0.4kg

Human frontal cortex 3-4 times bigger than chimpanzee's

Also, for reasons we still don't fully understand, there are a couple of mental processes our brains are spectacularly good at compared with all other species, including chimps.

A KNACK FOR EXPLAINING AND SHARING

1. **Seeking Explanations**. We humans don't just sit back and accept our fate, we always want to know **why** things happened. This curiosity is a fundamental part of being human. It's at the heart of most stories and all kinds of scientific research: we're always looking for ways of making sense of the world and our place in it.

2. **Teaching**. Some species of animal share their knowledge with each other to help their kind get ahead in life. Adult meerkats, for example, teach their young to deal with dangerous prey, like scorpions. They'll catch a scorpion, nip out its stinger and then pass it on to the youngsters to practise their hunting skills. But no other species sends its kids off to school, day in, day out, for years on end! And then basically goes on reading and watching and learning for the rest of their lives.

> Asking 'why' and teaching? Is that all you can come up with?

Well, both of these exceptional brain skills depend on our species' gift for mastering complex language. Put it all together, Octopus, and you can see why we human beings just keep on understanding more and more about the world. Every new generation has a chance to absorb everything humans everywhere on Earth knew before them. Then they can add their own ideas into the mix and pass those on to future humans.

> So you can go on and on making your lives better, without worrying about the havoc you cause for the rest of us! Fan-blooming-tastic.

You're right, Octopus. We are the species causing the climate **and** the *biodiversity crises*[*] and it's making life so much harder not just for ourselves, but for most wild species too.

You honestly still believe you're the smartest creatures on Earth?

Well, we might not always make the best decisions, but the thing is, until recently, most people didn't understand that our way of life was harming the entire planet. We do know now, so it's up to us to learn all we can about the problems we're causing, then use our unique intelligence to dream up solutions that we can share far and wide.

It's a huge task, but some people are optimistic that a totally new kind of brain, quite unlike any belonging to a human, animal or plant, might be emerging – just in time – to help us.

A new brain . . . what on Earth?

01010101101010101 **IS AI THE ANSWER?** 001001101101100

We're talking about *artificial intelligence* (or AI for short). You might already know that AIs are computer programs built to imitate the ways biological brains work. Most AI systems are made of 'virtual neurons' – basically lines of software code – that mimic the actual neurons we have in our brains.

[*]Find out more about these crises, and how we might fix them, in *Explodapedia: Rewild*.

The latest AI systems contain trillions of virtual neurons, linked up to form vast 'neural networks'. They can hoover up colossal amounts of information – including more books, photographs, videos and pieces of music than any individual person could absorb in a thousand lifetimes – and then use it to do incredibly brain-like things.

AI *chatbots* now 'understand' practically any question we ask them – and come up with detailed, creative and convincingly human-like answers. What's more, AI is already turbo-charging some areas of science research by, for example, discovering new medicines and improving renewable energy solutions for the climate crisis. AI might even help us finally understand just how smart other creatures are. A 2024 study, for example, showed that humans are not the only species to give each other names. African elephants do too – and they'll only respond if the members of the herd pronounce their name properly.

Actually it's Gnelly, with a silent 'G'.

But, as we're about to discover, some people think one of the biggest ways AIs could change our lives – for better or, possibly, for worse – is by communicating directly with our brains.

Towards a Brain-Bending Future?

Get ready to have your mind boggled, because we're about to leap forward through time to one possible version of the year 2075 . . .

Well, all is not quite as it seems. These are the students we met on pages 124-125, they've just been teleported into the future. And they're **not** bored, because tucked away inside each of their skulls is 2075's latest 'must-have' technological device – invented and operated with the help of cutting-edge AI, it's called a 'whole-brain interface'.

A what?

Remember the 'brain-computer interface' (BCI) on pages 52-53 that allowed Philip O'Keefe to control his computer by mind power? A whole-brain interface is like that, only a million times more powerful. Instead of reading information from one small patch of brain that controls movement, it connects to the **entire** cortex. As well as reading thoughts coming **out** of the brain, it can send ideas, feelings, memories, etc., **in**.

Here, with these headsets, you can see what's happening inside the students' minds. They're each on a 'learning journey' that's perfectly tailored to the needs of their unique brains.

Future Learning Journeys

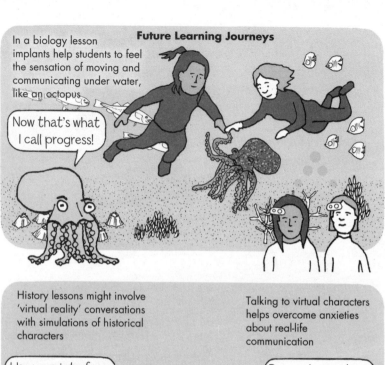

In a biology lesson implants help students to feel the sensation of moving and communicating under water, like an octopus

Now that's what I call progress!

History lessons might involve 'virtual reality' conversations with simulations of historical characters

Talking to virtual characters helps overcome anxieties about real-life communication

Use your interface to promote peace and harmony.

Inequality and injustice are everywhere – don't lose touch with reality.

Remember: actions speak louder than thoughts!

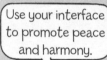

Mahatma Gandhi

Nelson Mandela

Emmeline Pankhurst

Dyslexia isn't an issue, since reading is now optional – and information on the internet can be searched for, and digested, just by thinking

Whole-brain interfaces mean ideas, memories and feelings can be shared directly, via telepathy

Installable 'mood modules' help students control emotions – mental health is generally excellent

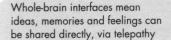

Skills, such as basic musical techniques, can be downloaded, but students still have to practise hard to keep up with the band

If attention wanders, interface gives a subtle prompt to regain focus

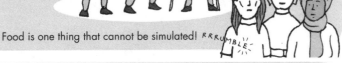

Food is one thing that cannot be simulated!

Future school looks totally brilliant! Can I stay here?

Nope, not an option. The time machine only lets us **witness** potential future scenes. We're not really here.

Can we just get on and build a future like this, then?

First things first. You've just seen one possible future, there's no guarantee things will pan out even remotely like this. Here in the 2020s we can already use BCIs with AI to make sense of signals coming **out** of the brain (see pp. 52-53). But to change someone's thoughts, feelings and perceptions, a whole-brain BCI would also need to turn neural circuits on and off by sending information **in** to the brain. We're already getting there. Today, thousands of blind people are already being helped by 'retinal implants' – fairly simple BCIs that use a camera to 'see':

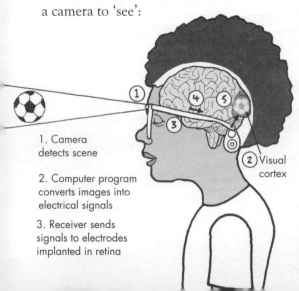

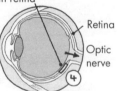

Close up of eye: Electrode implanted in retina

Retina

Optic nerve

1. Camera detects scene

2. Computer program converts images into electrical signals

3. Receiver sends signals to electrodes implanted in retina

2 Visual cortex

4. Retinal implant sends signals to optic nerve, which delivers information to visual cortex

5. Today's retinal implants provide the visual cortex with fairly limited vision

Neuroscientists would need all the decades between now and 2075 to develop whole-brain interface technology. Here's how it **could** potentially happen:

• **2030s.** Medical BCIs gradually become more powerful. Many more people start using BCIs to restore vision, hearing, movement and memory.

• **2040s.** BCIs can influence mood, emotion and reward circuits. They're used to treat depression, anxiety and addiction.

• **2050s/60s.** 'Neural dust' allows accurate 'mind reading' across the entire cortex – thoughts can be shared *telepathically*, like this:

BCIs provide perfect sight and hearing

BCI treating depression

Paralysed person walks with BCI-controlled 'exoskeleton'

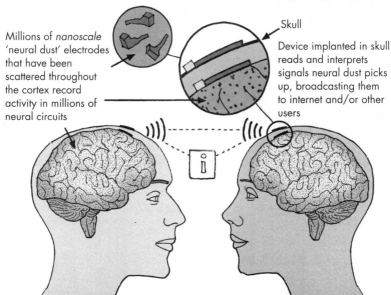

Millions of *nanoscale* 'neural dust' electrodes that have been scattered throughout the cortex record activity in millions of neural circuits

Skull

Device implanted in skull reads and interprets signals neural dust picks up, broadcasting them to internet and/or other users

• **2075.** The whole-brain interface shown at the start of this chapter might happen through advances in neural dust technology. More powerful electrodes could **activate**, and read from, neural circuits. **Any** circuit, **anywhere** in the cortex, could then be turned on or off **at any time**. Fully immersive, multi-sensory 'experiences' could be broadcast into people's brains, supercharging learning (see above), transforming mental health and offering amazing new forms of entertainment.

Bring it on!

Yes, but be careful what you wish for. Even if whole-brain interfaces worked perfectly and were completely safe – a big 'if', given how awesomely complicated the brain is! – there are lots of possible pitfalls. What if computer hackers managed to steal our most private thoughts, implant false beliefs or even control our entire bodies?

People might become addicted to constantly using their interfaces to trigger happy feelings. Maybe we need to experience life's lows in order to truly appreciate its highs.

What if some people became so entranced by their personal virtual worlds that they stopped caring for their bodies and lost interest in real-life experiences, including real people?

No way! I'd always want to hang out with my actual friends and family.

Well, you'd be able to choose the version of them that didn't make mistakes or annoy you.

But it wouldn't be **real**!

Is anything in your head really 'real'? Your brain never actually sees, smells or touches anything, remember? Even pain (see pp. 70-73) and colour (see pp. 73-76) are kind of illusions. So, if a whole-brain interface beamed **perfect** simulations into your brain, could you actually distinguish them from 'reality'?

You're all crackers! Isn't the world amazing enough as it is?

Yeah, this is all starting to sound a bit, er, creepy.

Thanks to neuroplasticity, our brains are brilliant at adapting to new experiences. But you're right to be cautious. We can't know how our minds would actually react to a BCI world of endless possibilities.

And this is just one potential future for the brain! As we said right at the start of the book, neuroscientists still don't know exactly how tiny worm brains work. **Our** brains contain 300 million times more cells. There's still so much we need to learn before anyone should even consider filling their brain with microscopic electrodes.

Besides, long before whole-brain BCIs are even an option, a different form of AI might take over the world.

Wait. What? 'AI takeover'!? You can't just casually drop that into the conversation!

Hmm, sorry, but it is an actual possibility that some AI experts are starting to worry about.

How Smart Can a Machine Be?

Right now, we're still very much in charge of telling them which problems to solve, but AIs are rapidly getting more powerful (see pp. 152-153). Looking further ahead, people could invent AIs that build and improve **themselves** – and even start setting their own intelligence goals. These 'artificial minds' would be able to learn and think **millions of times** faster than our squishy cell-based brains can.

Ha, maybe super-intelligent AIs of the future will treat you lot the same way you treat houseflies!

It's a possibility, Octopus. AIs won't necessarily turn out to be actively evil, but programming a super-intelligent AI only to do 'good things' could be surprisingly tricky. For example, give an all-powerful AI the goal of stopping the climate crisis and it might decide humans are the problem, then go all out to achieve its goal by devising an incredibly cunning way of killing us all.

Sounds more like super-stupidity!

Exactly. Intelligence is only useful when it's combined with good decision-making. Luckily for us, AIs are still nowhere near powerful enough to cause this kind of carnage. Even so the discoveries neuroscientists and computer scientists make

in the decades ahead have the potential to change our brains –
and therefore every aspect of our lives – in ways we can barely
even begin to imagine.

The Most Advanced Technology in the Universe?

For the time being, however, the world's most advanced
'technology' is still the one you carry around inside your
head. So perhaps the best way to approach the future is by
appreciating what we've already got.

Because, unlike you, even the world's most powerful AIs
still lack any hint of conscious awareness. And, unlike you,
they're not even the tiniest bit curious and they certainly don't
experience true empathy. It's humans who invent and build
AIs and other trailblazing technologies like BCIs. So we should
be the ones who choose how and when we use them.

If we want to avoid any nasty surprises – and, after all, that's
the main point of having a brain – we've got to tap into the
particular strengths of the amazing thinking, feeling, dreaming
cauliflowers inside our skulls. An essential part of that will
be working hard to understand much more about how brains
work.

Then, just as crucially, we've got to use all our new knowledge
to build a world where everyone gets a chance to flourish,
whatever challenges life throws their way and whatever kind of
brain they have.

That better include octopuses!

Of course it does! Because, as we now know, being a creature with a conscious, self-aware mind - octopus, human or otherwise - isn't always easy. A mind brings us grief as well as wonder. But it is also the most extraordinary gift. Imagine living your life without even a flicker of consciousness. It would hardly be like living at all.

Your brain doesn't just contain your universe. It is your universe.

GLOSSARY

Words italicized in this glossary have their own entries below.

action potential an *electrical impulse* sent by a *neuron* down an *axon*, away from its *cell body*

amygdala a region of your brain that processes emotions and certain kinds of memories

ancestor a group or individual from whom a living thing is directly descended

artificial intelligence a computer's ability to solve problems by learning and thinking

atom extremely small particles that bond together to build up all living and non-living things

autopilot a system that can control a vehicle without a human operator, often used in planes

axon a very thin cable-like part of a *neuron* that carries *electrical impulses* away from its *cell body* towards other *cells*

bacteria a tiny, single-*celled* living thing

basal ganglia part of the brain that mainly controls movements, but also helps you learn new skills, form habits and process emotions

biodiversity crisis damage to the natural environment that is causing species to go extinct and biodiversity (the variety of life found on Earth) to fall

brain-computer interface (BCI) a system that allows direct communication between the brain and an external device, like a computer

brainstem connects the brain to the spinal cord. It co-ordinates many of the brain's messages and controls several automatic bodily functions, for example, heart rate, digestion, breathing and swallowing

cell the smallest thing that can definitely be called 'alive'. Cells can live on their own as single cells, or together as parts of larger bodies

cell body the main part of a *neuron* that contains the *nucleus*

cell membrane a thin layer of *lipids* that surrounds a *cell* and most of its *organelles*

cerebellum a part of the brain in the back of the head, which helps with balance, co-ordination, and precise movements

cerebral cortex the outer layer of the brain, responsible for complex functions, including reasoning and processing language and information

chatbot an *artificial intelligence* computer programme that *simulates* human conversation through text and/or sound

cingulate cortex a part of the *limbic system* involved in emotions, decision-making and controlling attention

congenital a condition present at birth, rather than one which develops later in life

consciousness the state of being aware of one's self and/or surroundings

dementia a brain disease that kills off *neurons* faster than normal

dendrite part of a *neuron* which carries *electrical impulses* from other neurons to the *cell body*

dopamine a chemical released in the brain when we do something that makes us feel good (sometimes called the 'pleasure *hormone*')

dyscalculia a learning condition which makes it difficult for people to process numbers

dyslexia a learning condition which makes it difficult for people to read and write

echolocate the ability to work out where something is by using sound waves

electrical current the flow of tiny particles called electrons, which carry electricity

electrical impulse an electrical signal generated by *neurons,* which allows different parts of the *nervous system* to communicate

electrode a small piece of metal or other material that can conduct electricity, often used to measure brain activity

emergent property a feature that appears when many parts work together, but isn't found in any single part alone – e.g. an ant colony can only be built by a large number of ants

empathy understanding or being able to imagine how someone else is feeling

epilepsy a condition where damaged brain *cells* create abnormal *electrical impulses,* causing people to have fits

episodic memory memory of everyday events, including when and where something happened, who was there and how you felt

evolve/evolution the process by which living things gradually change over many generations

explicit memory memories you can consciously recall, such as facts, or events from your past

fovea part of the *retina* packed with 'cones' that detect colour and help you see in high-resolution

frontal cortex the front (and largest) part of the *cerebral cortex* which is involved in important brain functions such as planning, making decisions and personality

ganglia a large cluster of *neurons* and *glia*

gene each gene contains a specific instruction for how to build a particular part of a *cell* or body. Genes are passed down from one generation of a living thing to the next

glia *cells* in the brain that support, protect and help control *neurons*

glymphatic system a network of *glia cells* that spreads through the brain and helps remove waste

hemisphere one half of the brain; we have a left hemisphere and a right hemisphere

hippocampus part of the *limbic system*, it makes and recalls memories and helps you find your way around

hormone a chemical messenger that travels through the blood triggering activity in different parts of the body

hydra a tiny water animal with tentacles, that can regrow damaged body parts

hydrocephalus a condition where too much fluid builds up in the brain, causing damage

hypothalamus part of the brain that controls body temperature, thirst, hunger, sleepiness and emotions

immune cell part of the body's defence system that helps fight infections and other illnesses

implicit memory automatic memory that helps you remember how to do things like brush your teeth, tie your shoelaces or ride a bike

inflammation part of the body's protective response to illness or injury, involving *immune cells*, blood vessels and chemical messengers

intelligence the ability to use information to solve problems

interface a place where systems (or people) meet or work together

ion an *atom* or *molecule* that has an electrical charge

light-emitting diode (LED) a small lightbulb that uses very little electricity and doesn't get hot

limbic system a group of brain regions that deals with emotions, memory and behaviour

lipids oily *molecules* that are crucial for building all *cells*

magnetic field an invisible area of force around magnets or *electric currents*

magnetic resonance imaging (MRI) a procedure that uses a powerful *magnetic field*, radio waves and computer analysis to generate detailed images of the inside of the body

mental health your emotional, social and psychological wellbeing

mental-health condition a long-lasting problem that negatively affects your *mental health*, for example depression

microbe/microbial a form of life that is too small to be seen with the naked eye

microbiome a collection of different *microbes* that live in a particular place, such as the trillions of *microbial cells* that live on and in a human body

mind part of a person that thinks, feels, remembers, imagines and has *consciousness*

molecule two or more *atoms* joined together

motor cortex part of the brain that controls deliberate movements

motor neurons nerve *cells* that send signals from the *nervous system* to muscles to make them move

multicellular living things with bodies made from lots of *cells* that are working together

myelin a protective layer that forms round *nerves*, helping signals travel faster

nanoscale minute: one nanometre is one billionth of a metre

nerve *neurons* or bundles of neurons that carry electrical signals around the brain and body

nerve cord a thick bundle of *nerves* running through the body, from the head down the back

nerve net a simple *nervous system*

nervous system the network of *neurons, glia* and *nerves*, including your brain and spinal cord, that sends messages throughout your body

neural circuit a group of connected *neurons* that work together to achieve a particular task

neurodevelopmental to do with how the brain and *nervous system* grow and change as a person gets older

neurologist a doctor who studies and treats the *nervous system*

neuron a brain cell that carries *electrical impulses* around the brain and *nervous system*

neuroplasticity the brain's ability to change and adapt itself according to what's happening in and around it

neuroscience/neuroscientist the study of, or someone who studies, the *nervous system*

neurotransmitter a chemical that passes messages between *neurons*

neurotrophic something that helps *neurons* grow and survive

nucleus/nuclei the main control centre of the *cell* of a living thing, which contains the *genes*

nutrient a substance that helps the body and *cells* develop and grow

optic nerve the *nerve* that connects the eye to the brain

organelle a small structure inside a *cell* which does a specific job

perception how your brain interprets and organizes information from your senses

phobia a consistent, significant fear of a specific thing, animal or situation, for example, xanthophobia is a fear of the colour yellow

photoreceptor a *neuron* that converts light into *electrical impulses*

pupil the tiny opening at the front of your eye which controls how much light goes in to your eye

reflex an automatic reaction, e.g. pulling your hand away immediately from a hot surface

retina a layer of light-sensitive *photoreceptor cells* in the eye

reward circuit a system of linked neurons responsible for producing desire, cravings and pleasurable feelings when we do enjoyable things

semantic memories general knowledge and facts you've learned about the world, including the meanings of words, places and objects

sensitized when a *nervous system* becomes increasingly sensitive to or aware of a *stimulus* over time

sensory neuron *neurons* that receive information from your senses and send it to the *nervous system*

simulate to imitate or present something as real when it isn't

spectrum a wide range

stereotype an over-simplified belief about the characteristics shared by a group of people. Stereotypes are often unfair and untrue

stimuli/stimulus things that provoke a response or action

stroke when blood flow to part of the brain is blocked, causing brain *cells* to die

supercomputer a very powerful computer

synaesthesia a condition where senses are merged, for example, you might taste colours, or see music

synapse a tiny gap between *neurons* where messages are passed from one *cell* to another

synaptic pruning the process of removing *synapses* that aren't doing anything useful, as the brain learns and develops

telepathy/telepathically the ability to communicate thoughts directly from one *mind* to another

transcranial through or across the skull

unconscious brain processes that you are not aware of, but that affect your thoughts, feelings and behaviour

vagus nerve one of the longest *nerves*, running from the *brainstem* to the abdomen (stomach), which helps us digest food and control emotions

virtual reality a digital environment *simulated* by a computer

visual illusion a picture or video that tricks us into thinking something we see, which is false or impossible, is actually real or possible

visually impaired having reduced or a complete lack of vision

vocal cords two thin pieces of muscle that stretch across your voice box, and vibrate to make sounds when you speak

INDEX

A

action potentials 27, 29, 32, 48, 72
age, brain and 86–95
Aldini, Giovanni 41, 130
amygdala 22, 101
animals, studying 48–9, 121, 139,
    140–53
anxiety disorders 14, 128–9, 130, 132
artificial intelligence (AI) 152–3, 155,
    158, 162–3, 164
    Attention deficit hyperactivity
    disorder (ADHD) 126–7
Autism spectrum disorder (ASD) 126,
    127, 128
awe 137–9
axons 26, 27, 32, 45, 47

B

babies 13, 83, 86–8, 94–5
bacteria 55–6
basal ganglia 22, 101
bees, honey 13, 146–8
birds 145–6
bodily functions, basic 23, 107, 115
bodily needs 101, 102
bottom-up impressions 76, 77
brain-computer interface (BCI) 52–3,
    155, 158, 162, 164
brain damage/injuries 36, 43, 85, 92,
    93, 94–5
brain stimulation 130–1
Brain-derived neurotrophic factor
    (BDNF) 134
brain-to-body ratio 144–5, 150
brainstem 17–18, 32, 72, 101
breathing 17, 23, 119
Broca, Paul/Broca's area 43–4, 88

C

capacity, brain 94
cell membranes 21, 48
cells 21, 26, 44–5, 48, 59, 162
cerebellum 18–19, 88
Chalmers, David 112

Chittka, Lars 146–8
chromatophores 64
cingulate cortex 101
climate crisis 127, 152, 153, 163
colour 13, 109, 110, 161
cones 74, 75
consciousness 97–100, 103, 107,
    108–22, 164, 165
cortex (cerebral cortex) 19–20, 52,
    61–2, 72, 94–5, 103, 155, 159, 160
crows 145–6

D

debates, inner 7–9, 100–7, 115, 139
decision making 7–9, 91, 115–16, 118
default mode network (DMN) 138, 139
dementia 92
dendrites 26, 27
depression 14, 128, 130, 131, 132, 135
Descartes, René 38–41, 98, 110
Dickinson, Emily 10
diet, balanced 135
dissection 36, 38
dopamine 90, 102
dreams 97, 133
dyscalculia 127
dyslexia 127, 156

E

Eagleman, David 99–100
eating 17, 23
echolocation 142
Egyptians, ancient 34–5
electrical signals 27, 28, 44, 45, 47–8,
    111–12, 117, 149
electricity 41–2, 130–1
elephants 153
emergent property 113
emotions 14, 21, 22, 23, 37, 43, 65, 88,
    103, 132
empathy 90, 92, 108, 120–1, 164
energy 81, 119
epilepsy 111

Acknowledgements

The brain is a massive – perhaps even limitless – topic to cover in one little book. There's no way everything within these covers would have fit, let alone be such fun to read, if not for the inspiration and dedication poured into it by series editor Helen Greathead. Massive thanks to designer Alison Gadsby for bringing it all to life on the page, and to commissioning editor Anthony Hinton for support throughout. Thanks also to copy and proof editors Julia Bruce and Philip Thomas, to indexer Helen Peters and to the rest of team DFB, including Fraser, Bron, Phil, Rosie, Ruth, Meggie, Kate, Rachel, Kathy, Sadie, Liz and David. Special thanks to neuroscientist Emily Towner, for expert fact-checking and crucial insights from the frontiers of brain research.

About the Author and Illustrator

Ben Martynoga is a biologist and an award-winning writer. After a decade in the lab exploring the insides of brain cells, he swapped his white coat for a pen. Since then he has written about everything from the latest tech innovations to rewilding, running, stress, creativity, microbes and the history of science. He loves talking about science – and why it matters – with children and adults alike at science festivals, in classrooms or anywhere else. His writing appears in the *Guardian*, *New Statesman*, the *i*, the *Financial Times* and beyond. He lives, works, wanders and wonders (often all at once) in the Lake District.

Moose Allain is an artist, illustrator and prolific tweeter who lives and works in south-west England. He runs workshops and has published a book and an online guide encouraging children to draw, write and find inspiration when faced with a blank sheet of paper. Always on the lookout for interesting projects, his work has encompassed co-producing the video for the band Elbow's 'Lost Worker Bee' single and designing murals for a beauty salon in Mexico City – he's even been tempted to try his hand at stand-up comedy. His cartoons regularly feature in the UK's *Private Eye* magazine.